Health in Past Societies

Biocultural interpretations of human skeletal remains in archaeological contexts

Edited by

Helen Bush and Marek Zvelebil

BAR International Series 567

1991

Published in 2019 by
BAR Publishing, Oxford

BAR International Series 567

Health in Past Societies

ISBN 9780860547167 paperback
ISBN 9781407348643 e-book

DOI https://doi.org/10.30861/9780860547167

A catalogue record for this book is available from the British Library

This book is available at www.barpublishing.com

BAR Publishing is the trading name of British Archaeological Reports (Oxford) Ltd. British Archaeological Reports was first incorporated in 1974 to publish the BAR Series, International and British. In 1992 Hadrian Books Ltd became part of the BAR group. This volume was originally published by Tempvs Reparatvm in conjunction with British Archaeological Reports (Oxford) Ltd / Hadrian Books Ltd, the Series principal publisher, in 1991. This present volume is published by BAR Publishing, 2019.

BAR
PUBLISHING

BAR titles are available from:

BAR Publishing
122 Banbury Rd, Oxford, OX2 7BP, UK
EMAIL info@barpublishing.com
PHONE +44 (0)1865 310431
FAX +44 (0)1865 316916
www.barpublishing.com

CONTENTS

LIST OF CONTRIBUTORS

Helen Bush, Ph.D.
Department of Archaeology and Prehistory
University of Sheffield.

Cristina Cattaneo, M.A.
Department of Archaeology and Prehistory
University of Sheffield.

Keith Dobney, M.Sc.
Department of Archaeological Sciences
University of Bradford.

Leslie E. Eisenberg, Ph.D.
Department of Anthropology
New York University.

Alan Goodman, Ph.D.
School of Natural Science
Hampshire College
Amherst,
Massachusetts.

Anne Grauer, Ph.D.
Department of Anthropology
Loyola University of Chicago.

Simon Hillson, Ph.D.
Institute of Archaeology
University College London.

Michael Hyland, Ph.D.
Department of Psychology
Plymouth Polytechnic.

Christopher Meiklejohn, Ph.D.
Department of Anthropology
University of Winnipeg
Canada.

Win Scutt, B.A.
Plymouth City Museums
Plymouth.

Patty Stuart-Macadam, Ph.D.
Department of Anthropology
University of Toronto.

Marek Zvelebil, Ph.D.
Department of Archaeology and Prehistory
University of Sheffield.

LIST OF FIGURES

LIST OF TABLES

ACKNOWLEDGEMENTS

We thank Russell Adams, of J.R. Collis Publications, for preparation of the camera-ready copy of this volume. We also thank Mavis Torry and Eddie Moth for their help in the preparation of the text and illustrations.

We are grateful to our contributors for tolerating the protracted completion time of this volume, which arose from circumstances beyond our control.

PREFACE

The theme of this volume is health: its recognition in, and significance for, past societies. The volume has its origins in the session 'A Matter of Life and Death' which the authors organised as part of the Theoretical Archaeology Group (T.A.G.) Conference at the University of Sheffield in December 1988. The papers presented at that session are augmented here by four additional contributions.

The volume is divided into two halves, the first presenting concepts of health and introducing some of the ways in which the condition can be identified in past populations. The second half consists of five case studies, principally of European material.

Our introduction examines traditional approaches to the study of human skeletal remains from archaeological contexts, and compares them to the 'biocultural' approach which has developed in North America. The main features of this comparatively novel approach are presented and the significance of health at population level is considered. Following the introduction, the chapter by Helen Bush looks at definitions of health, and in particular the ecological definition upon which the biocultural approach rests. The chapter traces the progression from definitions of health to the interpretation of the condition in past populations through the medium of 'indicators of stress'. Using clinical studies as examples, Bush demonstrates that psychological factors are capable of affecting growth and proposes that they may thus influence the formation of 'stress' markers, since these reflect growth disturbance. The link between morbidity and psychological states is also made by Michael Hyland and Win Scutt; in Chapter Three they present a functional theory of illness to account for the relationship between these two factors.

Alan Goodman's chapter continues the exploration of the notion of health and of its significance for past societies. He demonstrates the connection between health and adaptation through an examination of the phenomenon of small body size.

In Chapter Five, Cristina Cattaneo explains how genetic factors influence disease susceptibility, and describes research aimed at retrieving molecular information from skeletal material in order that the genetic susceptibility to disease of a population can be considered when its aetiology under investigation. Part One is completed by Simon Hillson's contribution, which describes the utility of histological studies of ancient teeth.

The purpose of the five papers in Part Two is to demonstrate the biological approach and the application of concepts which were introduced in Part One. In Chapter Seven, Anne Grauer focuses on the controversial subject of palaeodemography. Her study of a large skeletal assemblage from medieval York combines archaeological, historical and palaeodemographic data in order to illustrate the pitfalls and potential advantages of this field of study.

The next two chapters each concentrate on a single skeletal indicator of health status. Hillson introduced the dental lesions known as enamel hypoplasia in Chapter Six, and in Chapter Eight Keith Dobney and Alan Goodman report on ethnographic studies of this lesion from two very different locations, Bradford and Mexico. Complexities are revealed which give the lie to over-simplified interpretations of enamel hypoplasia in archaeological and skeletal assemblages. In the following chapter, Patty Stuart-Macadam presents the results of her study of porotic hyperostosis (a condition indicative of anaemia) in the skeletal assemblage from Poundbury Camp, a Romano-British site in Dorset. The chapter includes a discussion of the complex aetiology of the condition.

In Chapter Ten, Leslie Eisenberg describes her biocultural analysis of the disease which she noted in a skeletal assemblage from Averbuch, a prehistoric village site in what is now Tennessee. Her chapter is a clear exposition of the way in which health status can be used to postulate causes of culture change.

The final chapter in the volume is by Christopher Meiklejohn and Marek Zvelebil, who set out to show how a biocultural analysis of skeletal remains can shed light on the transition from hunting and gathering to food production in Europe.

The introductory chapter explains our motivation in bringing together these contributions. The final sentence of the volume expresses our hopes for it: 'If (the volume) has succeeded in demonstrating the potential and the value of the biocultural approach, then it will have served its purpose'.

Helen Bush and Marek Zvelebil
Sheffield
February, 1991.

PART 1

Concepts and Methods

1. PATHOLOGY AND HEALTH IN PAST SOCIETIES: AN INTRODUCTION.

Helen Bush and Marek Zvelebil

Approaches to Palaeopathology and Health in British and American Practice

The skeletal remains of past populations have been a focus of interest for physical anthropologists and medical practitioners since the nineteenth century. Parallel lines of investigation existed at this time: physical anthropologists were chiefly concerned with metric analysis (Washburn, 1953), while the attention of other scientists was turned to pathological conditions. The term "palaeopathology" was coined independently in America and Europe, and it was Sir Marc Armand Ruffer in 1913 who, describing his study of disease in Egyptian mummies, defined palaeopathology as "the science of the diseases which can be demonstrated in human and animal remains of ancient times" (Moodie, 1923:21). The work of Ruffer and of others, such as Moodie and Elliot-Smith, was essentially classificatory (Ortner, 1988), the broader implications of disease going unconsidered. The elucidation of the natural history of specific diseases continues to be the goal of many students of skeletal remains:

> "...skeletons can yield useful information on the presence or absence and incidence of certain bone and joint diseases in our ancestors and can add valuable historical perspective to our studies of the various diseases afflicting mankind today" (Rogers and Dieppe, 1983:1194).

We would argue that this clinical emphasis goes some way to explaining the lack of interest of many British archaeologists in human skeletal remains; archaeologists have been actively discouraged from involvement in their study:

> "There has been a lot of confusion because many of those studying skeletal material have been archaeologists or anthropologists with no medical background so that significant bony change may be missed or textbook descriptions of disease misapplied (Rogers and Dieppe, 1983:1192).

> "...a profound clinical knowledge is needed to solve the problems that are met. It is most unwise for anthropologists who lack clinical training to venture into the infinitely subtle field of ancient disease" (Wells, 1964:20).

> "Although being a doctor is a necessary condition, the condition is not sufficient. It is necessary, as well, to get a good anthropological foundation including, if possible, a good archaeological background, not only to escape the pseudopathology pitfalls but, overall, to realize the necessary collaboration with archaeologists" (Dastugue, 1980:3).

The inference is that "the necessary collaboration with archaeologists" is a one-way process to the benefit of the palaeopathologist. It is small wonder that many archaeologists are indifferent, dismissive and sometimes even hostile towards human skeletons and their excavation.

A further explanation for such unfavourable attitudes may be that of ignorance. One of us recently spent a period of time excavating an urban cemetery, which presented an excellent opportunity to elicit the opinions of those actively engaged in excavation. One such, whose uncovering of an adolescent skeleton was meticulous, admitted to having no idea what its subsequent study could reveal. Others referred inevitably to the detection of disease, and to the assessment of age-at-death and gender.

Unaware of the potential of human skeletal remains, many archaeologists view them as, at best, an irrelevance, and when encountered *in situ* as objects whose excavation is time-consuming and which somehow does not constitute "real" archaeology. Implicitly or explicitly, skeletal remains are held to be inferior to other artefacts, even though the folly of paying more attention to the fibula which fastens the shroud than to the corpse itself has been exposed (Dennell, 1979). There are, of course, numerous British archaeologists with extensive experience of skeletal excavation and informed enthusiasm for their analysis. Even so, we were amused by one who unconsciously betrayed himself when recounting the theft of skulls from the site which he was directing: fortunately, he remarked, none of the artefacts was stolen!

The limited concept of human skeletons may be attributed to the way in which they have been reported in the past, as appendices to excavation reports, rather than as an integral part of the whole. It may also be a consequence of the publication of new modes of skeletal analyses in specialist journals with which archaeologists are unacquainted. Yet one more reason may be the general tendency to describe all skeletal analyses as "palaeopathology". This is too narrow a description; it gives a false impression of what is, in fact, a broad field of study. Indeed, almost twenty years ago, Brothwell (1972) emphasised the minor role which palaeopathology plays in the reconstruction of the health of past populations.

More recently, considerable efforts have been made to promote the multi-faceted nature of burial archaeology in general (Roberts *et al*, 1989) and skeletal analysis in particular. In 1987, for the first time, a session dedicated to the study of human skeletal remains was organised as part of the Theoretical Archaeology Group conference, and when in the following year T.A.G. came to Sheffield, we welcomed the opportunity to build upon this precedent. The theme of our session was in deliberate contrast to that of the previous year; although we applauded the inclusion of skeletal studies in a theoretical conference, we felt that the tone of the papers, with the exception of that by Hodges, was still that of traditional palaeopathology, and we believed that the interest of archaeologists might be stimulated by a series of papers which demonstrated the "biocultural" approach to skeletal analyses. Indeed, we believe this to be true not just for our conference session, but also for European skeletal studies in general.

The "biocultural" approach rejects the traditional clinical mode of skeletal study which focuses on the individual, aiming instead at comprehension of the population to which that individual belonged. The two approaches are distinguished further in that rather than elucidating the natural history of specific disease, the biocultural approach is directed to answering questions about the *general health* of a population, frequently through the analysis of a range of skeletal and dental features, which are often referred to as "indicators of stress". These include the assessment of adult stature and of sub-adult growth rates; dental crowding; defects of the dental enamel; osteoporosis and skeletal lesions indicative of anaemia. Their interpretation was initiated in response to criticisms by "new" anthropologists of traditional skeletal studies (Buikstra and Cook, 1980). Health status influences life expectancy, reproductive potential and the capacity for labour (see Goodman, this volume), so as Brothwell (1972:359) has rightly stated, "a consideration of community health has a very real relevance to the proper study of early societies". We do not suggest that the identification of specific diseases has *no* place in the reconstruction of a society's health; the papers in this volume by Eisenberg, and by Meilkejohn and Zvelebil are testimony to its value in this respect. Nor, indeed, should disease and health be divorced from each other conceptually (see Bush, this volume). It is rather that the archaeologist, concerned with the "illumination of human society and its dynamics" (Ackernecht, 1953:126), seeks a broader view than a focus on disease *per se* can provide. According to Ortner (1988) this is now being recognised increasingly by palaeopathologists who have begun to inquire into the significance of the diseases which they detect skeletally, rather than simply attempting their diagnosis. So long as the latter remains the goal, the archaeological potential of human skeletal remains is wasted.

Inevitably, there are difficulties and disagreements when skeletal remains are analysed in this way. The wider application and evaluation of indicators of stress has revealed that in some cases their interpretation in terms of the health of past populations has been naive. A classic example is that of the transverse lines which are frequently observed on the radiographs of long bones, and which are commonly referred to as Harris lines. These were heralded as "a new approach to ancient disease" (Wells, 1961:526). With his customary eloquence, Wells (1967:390) summed up the picture which traditional physical anthropology paints for us: "...we can see an early individual at only one moment - the moment of death. It is a snap-shot, not a moving picture". It was in order to move from this static position that he promoted the radiological examination of archaeological skeletons for the presence of Harris lines; they would be a step towards a diachronic portrayal of the life of an individual, because it was believed that they reflect health status at different stages during the years of growth, as opposed to health at the moment of death, if such a state can be said to exist!

Wells used Harris lines to make social inferences. For example, from an inter-populational study in which the relationship between gender, age-at-death and Harris line frequency was assessed, he proposed that the greater number of lines in females indicated preferential treatment of males from infancy. Women, he concluded, were "the undernourished drudges of their menfolk" (1967:398). Although subsequent research (e.g. Garn and Braunstein, 1986; Mensforth, 1981; Murchison *et al*, 1984) has cast doubt upon the validity of Harris lines as measures of health, Wells' recognition of the need to move away from traditional approaches was pioneering.

Concern to reveal the anthropological relevance of osseous remains has resulted in the utilisation of skeletal and dental features whose significance in terms of health status are not clearly understood. Harris lines are a clear example; dental enamel hypoplasia (a defect produced by a slowing or cessation of enamel production during the initial stage of enamel formation) is another. Ethnographic studies are essential to our greater understanding of such lesions and thus to more accurate reconstructions of the health of past societies.

It will be apparent from this volume that there is no consensus of opinion between researchers with regard to the measures of health which they utilise. Stuart-Macadam, for example, reiterates her previously (1982; 1985) expressed view that porotic hyperostosis reflects anaemia experienced in childhood; conversely, Eisenberg reports active lesions in adult crania. The debate concerning the age at which anaemic episodes occurred hinges on the differentiation of remodelled from unremodelled lesions; currently there are no generally recognised standards on which to base such discrimination.

Palaeodemography as a branch of physical anthropology is particularly controversial, being virtually dismissed by Bocquet-Appel and Masset (1982) but stoutly defended by, for example, Buikstra and Konigsberg (1985) and Van Gerven and Armelagos (1983). Recent work by Suchey and colleagues (Suchey & Katz, 1986; Suchey *et al*, 1986; Suchey *et al*, 1988) on the estimation of age-at-death through morphological changes of the pubic symphyseal face has revealed major discrepancies in similar, frequently-used techniques, and these discrepancies have been confirmed by the study of individuals of known age-at-death from Christ Church, Spitalfields, (Cox, pers. comm.). These revelations have serious implications for skeletal analyses, since estimations of age-at-death are fundamental to any attempt to reconstruct the health of past societies. Nor are the uncertainties of ageing peculiar to the adult segment of a skeletal assemblage.

The Biocultural Approach in this Volume

In comparison to traditional approaches to human skeletal remains, the biocultural approach is a new area, and as with any new discipline, over-simplification and variations in interpretation are to be expected, particularly in one which incorporates as many unknown variables as does archaeology. The recognition of these problems should be viewed in a positive light since it stimulates further research, while introducing realism to skeletal studies. In contrast to North America, the biocultural approach is yet to become established in Europe, but although population analyses are lacking, with the exception of St. Helens-on-the-Walls (Grauer, 1989), theoretical and methodological studies of measures of health are more abundant (e.g. Bullion, 1987; Bush, 1990; Dobney, 1983, forthcoming; Stuart-Macadam, 1982). The purpose of the T.A.G. conference session which we organised in 1988, and its translation, with additional papers, into this volume, was to illustrate research inspired by biocultural concepts on both sides of the Atlantic and promote such approaches in the archaeological community at large.

What, then, are the main features of the biocultural approach? In essence, the biocultural approach can be seen as an elaboration of concepts and relationships developed in cultural ecology and applied within the framework of the evolutionary theory. In much of the earlier work striving to elucidate the dynamic relationships between population, culture and the environment, the condition of the *actors* - individuals, social groups and populations - has been neglected, either due to the lack of paradigmatic awareness, or because of poor evidence. In trying to understand the relationship between population and culture, the aim of the biocultural approach is twofold: (1) to assess the biological condition of human populations and its consequences for the biological and cultural reproduction of the society and (2) to consider the selective effects of culture on the population under study and its survival. Unlike sociobiology, where genetic interests are seen as pivotal in shaping the basic structure of society, the biocultural approach places emphasis on the integration and inter-relatedness of biological and cultural systems, without according causal priority to either.

In pursuing this approach, we have to outline a few key concepts and relationships, despite an inevitable divergence of views held about some of the complex phenomena discussed briefly below:

1) The human organism, and its surviving remains, is comprehended as a dynamic, historical and adaptive system, reflecting the interaction between the biological organism and its environment during the individual's lifetime. Health is the biological and psychological condition of the individual, a state which can, at least in theory, be assessed by the incidence of insults and other abnormal symptoms. The two major problems which arise early in such assessment are the multiple causes which can result in the same type of insult, and the hierarchical order in which insults are expressed, the skeleton being one of the last systems to be affected. The first problem means that a symptom is not usually related to a single cause or stressor: this is well illustrated by Dobney and Goodman in their paper on enamel hypoplasia, and justifies the need to use multiple stress indicators (including archaeological evidence) in evaluating the potential causes of stress. The second problem means that some insults, if acute and/or system specific, may not leave any evidence of a skeletal response (see Goodman, this volume). The individual may die before the skeletal response develops, or the insult may be recovered from without leaving any traces.

Health is evaluated by cultural perceptions and parameters: ill-health and disease are culturally defined phenomena. In this non-biological sense, the definition of health can range from one of physical and mental well-being, with anything else being ill-health, to the clinical presence of disease indicators or of the biological response to it (see Hyland and Scutt's discussion of the "medical model") with anything other considered healthy. This serves to illustrate that the concept of health itself is a problematic and culture-specific notion, and points to the differences between cultural and medical/clinical perceptions of health (see Bush, this volume). As students of past societies, we are interested in both perceptions of health, but only rarely and by reference to the not-too-distant past are we able to understand the cultural definition. This leaves us with the clinical assessment of health as the only cross-cultural way of evaluating the health status of past populations from the archaeological record, yet as Bush describes in this volume, clinical definitions of health cannot provide a comprehensive framework for describing the condition.

2) Since humans are social creatures who spend most of their lives in the company of others, the health of individuals cannot be assessed fully without evaluating the health of other individuals within a community. Equally, since humans depend for survival on co-operation or competition with others and since the passage of social traditions is subject to control at household, community and broader population levels, only the study of health and demographic parameters at these levels can answer questions about the influence of health on biological and social reproduction. Within the biocultural approach, therefore, the emphasis is placed on the health condition of communities and populations, as well as individuals. The concept of community health and its applications are developed further by Eisenberg in this volume.

3) The key objective of the biocultural approach is to understand the role of health in the dynamic interaction between the environment, human populations and culture. Such an objective addresses key concepts in the study of human cultural and biological evolution, such as those of adaptation and fitness. As Goodman *et al* noted in an earlier publication:

> "... stress is a potentially significant force for change in fitness of cultural and biological systems, a topic of central concern among anthropologists" (Goodman *et al*, 1980:516).

These concepts and relationships are not easy to define and difficult to illustrate without gross generalisations. In archaeology, we have only vague and discordant notions of the meaning and role of such processes as natural selection as opposed to cultural selection and adaptation, and of such states as fitness (e.g. Dunnell, 1980; Ingold, 1986; Mithen, 1989; Rindos, 1984, 1985; Redding, 1988, etc.). And yet the biocultural approach aims at elucidating the role of these fundamental forces in the continuing story of human existence.

Even a moderately comprehensive survey of these concepts is beyond the scope of the present volume, and deserves a tome in its own right. But since these concepts are so germane to the approach adopted in this volume, an operational and heuristic characterisation is offered below:

a) There are dynamic links between the natural environment, the social environment and human biology. Change occurs through selection (whether cultural or natural) acting on the variability in human behaviour and human biology, selecting for those traits which are adaptive in the relationship to the social and natural environment of the time (see also the contributions to this volume by Eisenberg and Goodman).

b) Adaptiveness can be defined in terms of inclusive Darwinian fitness: the greater chances of some individuals to survive and reproduce, relative to others, in a given social and environmental situation (but see Hyland and Scutt, this volume).

c) Health and disease are conditions affecting the chances of individuals to survive and reproduce, and, by extension, the reproduction of the social environment. Health and disease, then, are major elements of selection.

d) At the same time, health and disease are products of human biological variability (in the genotype and the phenotype), and the selective pressures imposed on it by the social and natural environment. Behaviour which fails to resolve these pressures to the benefit of the individual "can result in impaired physical and psychological well-being, in disease and ultimately in death" (Cox, 1983:255). The condition of health, then, can be seen also as a reflection of the human biological adaptation to its environment (see Goodman, this volume).

A number of papers address these issues in the present volume. Eisenberg, in her study of the Middle Cumberland groups in Tennesee, considers the implications of dietary stress as well as *social stress* (see also Johnson, 1982, for the notion of *scalar stress*), both products of the *social environment* as well as natural conditions. Archaeological, cultural data play a crucial role in identifying the constraints imposed by the social environment, and Eisenberg's interpretation clearly would not be as complete had either the biological or cultural data been omitted.

Goodman presents a model in which health is a central feature of the adaptive process, and considers the position of health in promoting or hindering cultural and social reproduction. The contributions of Cattaneo and of Hyland and Scutt address the question of genetic as opposed to environmental factors in the health status of individuals and populations. Hyland and Scutt go beyond this controversy and posit the contradiction between individual survival and genetic evolution. Adopting trout as an analogue, they define biological adaptation as the continuation of the gene pool, and emphasise reproduction as the key element in the definition of fitness. This may be at the expense of survival: so long as the individuals survive long enough to pass on their genetic inheritance, they may be deemed fit. This is thought-provoking, but as Hyland and Scutt point out, the increased longevity of modern humans, well beyond their prime reproductive years, may be seen as a cultural adaptation to educate and pass on social traditions from one generation to another, thereby increasing the survival and reproductive chances of the offspring. Natural selection, therefore, would select for culture and the growth of social traditions, *so long as it confers selective advantage on culture-conscious groups in relation to others* (our emphasis). The implications of the relative nature of the survival value of culture are yet to be explored.

Hyland and Scutt's paper casts a relative value on palaeodemographic measures such as the life-expectancy, survivorship, morbidity and mortality of a population as the ultimate and cross-cultural measures of adaptive fitness.

4) At a methodological level, two major problem areas exist. First how far is the burial/mortuary/mortality sample representative of the conditions in the living population from which it was taken, and second, how representative of individual health status are skeletal and dental markers of insult? The very fact that archaeological skeletal assemblages *are* mortality samples means that they are unlikely to be a direct reflection of a living population (Cook, 1981). We have to be aware of a range of biasing factors:

a) social traditions and ideology mediate between the living and the dead and act as a selective code as to where and how people are buried, and who is and who is not treated thus.

b) processes of differential preservation and alteration act upon the buried human remains.

c) a number of factors ensure that the frequency of markers of insult in a skeletal population will not reflect that of the living population from which it was derived. Their frequency will be affected by the interaction of (1) the age structure of the indicator in the living population and age-specific death rates and (2) causes of death and the frequency of the indicator (Cook, 1981). It is common practice to compare skeletal assemblages which differ chronologically or geographically, yet Cook (*ibid.*) has shown that the degree of growth disturbance, and by extension, the frequency of markers of insult, is unlikely to be comparable between two discrete populations in which the causes of death are different.

The degree to which the health status of an individual is reflected skeletally depends to a certain extent upon the nature of the insult which is encountered. Some insults will not affect the skeletal and dental systems, while others may be so acute that the individual dies before these systems are involved. In connection with this, the point made forcefully by Goodman in this volume is worth remembering: markers of insult are signs of a struggle to adapt. It has been assumed in the past that a greater number of markers of insult means a greater degree of ill-health, but it may be that the opposite is true, for not only are such markers testimony of an encounter with stress, but also of rallying from it. This is well-documented by the example of severe infection: such a condition would make heavy demands upon the immune system, with a consequent lowering of chances of survival (Lallo *et al*, 1978), yet paradoxically, since bone is generally the last system of the body to be affected by infection, the immune system has to be able to respond sufficiently for the skeleton to be affected (Ortner, 1988).

A further major point to consider concerning the frequency of markers of insult is that the response to the "stress" which produced them is not uniform either within or between individuals (see Bush, this volume). Thus there may be temporal variation in resistance, stress being experienced only when resistance is lowered. This leads to a problem of interpretation: absence of a marker of insult may be indicative of heightened resistance, but conversely it may mean failure to encounter a stress-invoking situation. Moreover, variation of response between individuals may be a circumstance of genetic factors, age and gender (Goodman *et al*, 1984). The relative frequency of skeletal markers of insult in males and females is routinely assessed in archaeological assemblages and frequently interpreted in cultural terms; Wells' use of Harris lines, to take just one example, was referred to earlier. In the face of such cultural interpretations of markers of insult, it is noteworthy that women may be hypoactive with regard to their sympathetic-adrenomedullary response (Cox, 1978), for this may well influence the formation of markers in women. So too may the greater resistance of girls than boys to the consequences of illness and malnutrition (Tanner, 1977). Variation in the frequencies of markers of insult in women as compared to men may be a reflection, not of cultural practices, but of differing physiological activity.

As this short introduction shows, the study of health in past societies, from the reconstruction of basic parameters to the assessment of wider implications, is a very complex field in the early stages of practice. In our view, however, health as a measure of fitness, and as a condition influencing social and biological reproduction, is central to our understanding of human biological and cultural development. It is only at the interface of archaeology and cultural and physical anthropology, within the framework of the biocultural approach, that the study of health in past societies can be fully developed.

References

Ackerknecht, E.H. (1953). Paleopathology. In *Anthropology Today*, ed. A.L. Kroeber. pp. 120-26. Chicago: University of Chicago Press.

Bocquet-Appel, J-P. and Masset, C. (1982). Farewell to paleodemography. *J. Hum. Evol.* 11:321-33.

Brothwell, D.R. (1972). Community health as a factor in urban cultural evolution. In *Man, Settlement and Urbanism*, eds. P.J. Ucko, R. Tringham and G.W. Dimbleby. pp. 353-61. London: Duckworth.

Buikstra, J.E. and Cook, D.C. (1980). Palaeopathology: an American account. *Ann. Rev. Anthropol.* 9:433-470.

Buikstra, J.E. and Konisberg, L.W. (1985). Paleodemography: critiques and controversies. *Am. Anthropol. 87*:316-33.

Bullion, S. (1987). *Incremental Structures of Enamel and their Applications to Archaeology.* Unpublished Ph.D. thesis, University of Lancaster.

Bush, H.M. (1990). *The Recognition of Physiological Stress in Human Skeletal Remains: a Critique of Method and Theory with Specific Reference to the Vertebral Column.* Unpublished Ph.D. thesis, University of Sheffield.

Cook, D.C. (1981). Mortality, age-structure and status in the interpretation of stress indicators in prehistoric skeletons: a dental example from the Lower Illinois Valley. In *The Archaeology of Death.* eds. R. Chapman, I. Kinnes & K. Randsborg. pp. 133-44. Cambridge: Cambridge University Press.

Cox, T. (1978). *Stress.* London: MacMillan Press Ltd.

Dastugue, J. (1980). Possibilities, limitations and prospects of paleopathology of the human skeleton. *J. Hum. Evol.* 9:3-8.

Dennell, R. W. (1979). Prehistoric diet and nutrition: some food for thought. *World Archaeol.* 11:121-35.

Dobney, K. (1983). *Enamel Hypoplasia of Pig Teeth: its Relevance and Application to Zooarchaeological Studies.* M.Sc. Thesis, University of London.

Dobney, K. (forthcoming) *An Epidemiological Study of Dental Enamel Hypoplasia in Bradford School Children: its Relevance to Archaeological Studies.* Ph.D. thesis, University of Bradford.

Dunnell, R.C. (1980). Evolutionary Theory and Archaeology. In *Advances in Archaeological Method and Theory* Vol. 3. ed. M. Schiffer. pp. 38-99. Orlando: Academic Press.

Garn, S.M. and Braunstein, E.M. (1986). *Contributions of multi-imaging techniques to an understanding of structures represented by radiopaque lines and bands in tubular bones.* Paper presented at 55th Annual Meeting of American Association of Physical Anthropologists, Albuquerque, New Mexico, April 9-12.

Goodman, A.H., Armelagos, G.J. & Rose, J.C. (1984). The chronological distribution of enamel hypoplasias from prehistoric Dickson Mounds populations. *Am. J. Phys. Anthropol.* 65:259-66.

Grauer, A.G. (1989). *Health, Disease and Status in Medieval York.* Unpublished Ph.D. thesis, University of Massachussetts, Amherst.

Ingold, T. (1986). *Evolution and Social Life* Cambridge: Cambridge University Press.

Johnson, G. (1982). Organizational structure and scalar stress. In *Theory and Explanation in Archaeology.* eds. C. Renfrew, M. Rowlands & B. Seagraves. pp. 389-41. Orlando: Academic Press.

Mensforth, R.P. (1981). *Growth velocity and chondroblastic stability as major factors influencing the pathogenesis and epidemiological distribution of growth arrest lines.* Paper presented at Fiftieth Annual Meeting of the American Association of Physical Anthropologists, Detroit.

Mithen, S. (1989). Evolutionary theory and post-processual archaeology. *Antiquity 63*:483-94.

Moodie, R.L. (1923). *Palaeopathology: an introduction to the study of ancient evidences of disease.* Urbana, Illinois: University of Illinois Press.

Murchison, M.A., Owsley, D.W. & Riopelle, A.J. (1984). Transverse line formation in protein-deprived rhesus monkeys. *Hum. Biol.* 56:173-182.

Ortner, D. J. (1988). *Theoretical and Methodological Issues in Paleopathology.* Paper prepared in advance for participants in the International Congress of Anthropological and Ethnological Sciences Symposium on Human Paleopathology: Current Syntheses and Future Options.

Redding, R.W. (1988). A general explanation of subsistence change: from hunting and gathering to food production. *J. Anthropol. Archaeol.* :56-97.

Rindos, D. (1984). *The Origin of Agricultural Systems: An Evolutionary Perspective.* Orlando: Academic Press.

Rindos, D. (1985). Darwinian selection, symbolic variation and the evolution of culture. *Curr. Anthropol. 26*:65-88.

Roberts, C.A., Lee, F. & Bintliff, J. (1989). eds. *Burial Archaeology: Current Research, Methods and Developments.* British Archaeological Reports, British Series 211.

Rogers, A.R. (1988). Does biology constrain culture? *Am. Anthropol. 90*:819-31.

Rogers, J. and Dieppe, P. (1983). Lessons from Palaeopathology. *The Practitioner* 227:1191-99.

Stuart-Macadam, P. (1982). *A Correlative Study of a Palaeopathology of the Skull.* Unpublished Ph.D. thesis, University of Cambridge.

Stuart-Macadam, P. (1985). Porotic hyperostosis: representative of a childhood condition. *Am. J. Phys. Anthropol.* 66:391-8.

Suchey, J.M. and Katz, D. (1986). *Skeletal Age Standards Derived from an Extensive Multiracial Sample of Modern Americans.* Paper presented at the Fifty Fifth Annual Meeting of the American Association of Physical Anthropologists, Albuquerque, New Mexico, April 1986.

Suchey, J.M., Brookes, S.T. & Katz, D. (1988). *Instructional materials accompanying female pubic symphyseal models of the Suchey-Brookes system.* Distributed by France Casting.

Suchey, J.M., Wisely, D.V. & Katz, D. (1986). Evaluation of the Todd and McKern-Stewart methods for ageing the male *os pubis.* In *Forensic Osteology: Advances in the Identification of Human Remains.* ed. K. Reichs. pp. 33-67. Springfield, Illinois: Charles C. Thomas.

Tanner, J.M. (1977). Physical and psychological development. In *Human Biology.* 2nd ed. eds. G.A. Harrison, J.S. Weiner, J.M. Tanner & N.A. Barnicot. pp. 352-360. Oxford: Oxford University Press.

Van Gerven, D.P. and Armelagos, G.J. (1983). "Farewell to Paleodemography?" Rumours of its death have been greatly exaggerated. *J. Hum. Evol.* 12:353-60.

Washburn, S.L. (1953). The strategy of physical anthropology. In *Anthropology Today*, ed. A.L. Kroeber, pp. 714-27. Chicago: University of Chicago Press.
Wells, C. (1961). A new approach to ancient disease. *Discovery 22*:526-31.
Wells, C. (1964). *Bones, Bodies and Disease.* London: Thames and Hudson.
Wells, C. (1967). A new approach to palaeopathology: Harris' lines. In *Diseases in Antiquity* eds. D.R. Brothwell and A.T. Sandison. pp. 390-404. Springfield: Thomas.

Acknowledgement

We thank Dr. Margaret Cox for unpublished data relating to the skeletal remains from Christ Church, Spitalfields.

2. CONCEPTS OF HEALTH AND STRESS

Helen Bush

Introduction

Ecological definitions of health emphasise the dynamic nature of the condition, and in particular the ability of the individual to recover from insult. Increasingly, anthropologists evaluate the health of past populations, represented by skeletal assemblages, through the interpretation of dental and skeletal features caused by the human organism's adaptive response. These markers are commonly known as "indicators of stress".

This chapter explores medical and anthropological concepts of stress, and proposes that the adoption of Selye's physiological stress model by anthropologists has meant that psychological influences on the formation of stress markers have been ignored, by and large. The chapter presents examples from the fields of medicine and pyschology to demonstrate the impact which adverse mental states may have upon the growth process, and it is concluded that it is essential to keep in mind psychological factors when interpreting indicators of stress.

Definitions

The biocultural approach to human skeletal remains views them as remnants of a system which was involved reciprocally with its natural and social environment. Attempts to evaluate the interaction of these systems are frequently made through analyses of the health of past populations, because health status influences factors such as life expectancy, reproductive potential and the capacity for work and learning, all of which are crucial to the maintenance of a society.

From earliest times, until the introduction of the "germ" theory of disease by Pasteur and others in the late nineteenth century, disease was viewed as a disturbance of the balance between an individual and his environment, both internal and external (Dubos, 1959). Hippocrates, for example, described "the influence of climate, water supply and situation on health" (Chadwick and Mann, 1950:90). With the recognition of the link between disease and bacteria, the latter were considered to be the sole cause of the former (Cassel *et al*, 1960; Wylie, 1970). So, for example, tuberculosis was seen as a consequence of exposure to the tubercle bacillus, the symptoms of the disease being caused by disruption to the physiological system; perturbations of the psychological or social systems were not entertained as possible causes of the disease (Cassel *et al*, 1960). In recent years there has been a revival of the ecological concept of health, with a resurgence of interest in homeopathy and other holistic, "alternative" forms of medicine.

The ecological definition of health, which sees health and disease as successful and unsuccessful attempts to adapt to stress from internal and external sources, underlies the biocultural analysis of skeletal remains. It emphasises the dynamic nature of the condition, specifically the notion of adjustment to, or rallying from, unfavourable influences. For example:

> "Health is a continuing property, potentially measurable by the individual's ability to rally from insults, whether chemical, physical, infectious, psychological or social" (Audy, 1971:142).

> "Health is a function of the organism's ability to adjust to environmental constraints and stressors" (Goodman, this volume).

> "States of health or disease are the expression of the success or failure experienced by the organism in its efforts to respond adaptively to environmental challenges" (Dubos, 1965:xvii).

> "Health is the perfect, continuing adjustment of an organism to its environment" (Wylie, 1970:103).

The description of health as "continuing" reinforces the notion that the condition is one end of a spectrum which also incorporates disease. Health "does not disappear during an illness to return on recovery but continues, even though it may drop in level while the organism is adapting to the current insult" (Audy, 1971:142). According to Wylie (1970), the principal shortcoming of the contrasting definition of health, offered perhaps most famously by the World Health Organisation, is its inference that health and disease are discrete and mutually exclusive conditions:

> "Health is a state of complete physical, mental and social well-being and not merely the absence of disease or infirmity" (Preamble to the Constitution of the World Health Organisation, 1946, quoted by Wylie, 1970:101).

Wylie described this definition as "asymptotic", because it views health as a goal which ultimately cannot be achieved.

How do we progress from ecological definitions of health to the recognition and evaluation of the condition in the skeletal remains of extinct populations? The vehicle is the physical traces of the human organism's adaptive response, which are generally described as indicators of stress, or, less frequently but preferably, markers of insult.

This introduces a fundamental question: what is stress? A survey of the anthropological literature revealed a wide and sometimes undefined use of the term in reference to past populations (Table 2.1), which is significant since the concept of stress which has been the major influence in anthropological usage, namely that of Selye (1950; 1957; 1973), involved a clear definition as will be seen shortly. The clinical literature reveals that problems in defining stress are not, however, peculiar to anthropology. The popular use of "stress" is in the sense of nervous or emotional strain, and there are currently numerous books and magazine articles dedicated to advising business executives how to avoid its deleterious consequences. Mason (1975) noted that in the early years of this century clinical researchers, notably Cannon (1914; 1928; 1929), also regarded stress in this light. Although Cannon considered stress in a physiological context as well, particularly with regard to internal homeostasis, it is the work of Hans Selye which dominated this field for two decades after Cannon until the early 1950s, and which has been adopted by anthropologists to identify episodes of metabolic imbalance. Selye's pioneering work led to the formulation of a concept of stress, and the revelation of its physiological mechanisms (Cox *et al*, 1983). His theory of stress is notable for its three basic premises: that stress is a *response* to stimuli, that the response is *non-specific*, and that the manifestation of the response forms a recognisable syndrome, the General Adaptation Syndrome (G.A.S.). Although in a major work published in 1946, Selye used the term "stress" "in the conventional sense of stimuli, evocative agents, or outside forces acting on the organism" (Mason, 1975:9), his later work is characterised by his perception of it as a response to stimuli. The purpose of the response is the maintenance of homeostasis, or balance, of the system. At this point it will be helpful to consider the mechanisms by which this is achieved.

Physiological Considerations

It appears that the physiological response of the body is principally that of activity within the autonomic nervous system[1] and the endocrine[2] system (Cox *et al*, 1983). Under the influence of work by Cannon and Selye, attention has focused on the pituitary-adrenal cortical system and the sympathetic-adrenal medullary system (*ibid.*).

There are two adrenal glands, one lying over each kidney, and they each contain two endocrine organs, the outer cortex, which secretes steroids, and the inner medulla, which secretes the catecholamines (adrenaline and noradrenaline). The glands are well-supplied with nerves from the sympathetic system. The adrenal cortex produces three groups of steroid hormones: glucocorticoids, mineralocorticoids and androgens; the emphasis in stress research has been on the glucocorticoids (they form the greatest part of cortical secretion) although all three groups function in response to stress. Eightly five per cent of glucocorticoid secretion in humans is hydrocortisone, the remaining 15% being chiefly corticosterone (Cox *et al*, 1983). Glucocorticoids, which are essential to life, affect the metabolism of proteins, fats and carbohydrates, in particular the conversion of proteins into glucose and its storage as glycogen in the liver; they also affect the growth of connective tissue. In higher concentrations, as occur in clinical intervention or as the result of stress, glucocorticoids inhibit protein synthesis, block the inflammatory response, elevate blood sugar levels and initiate the loss of calcium and phosphate from the kidneys (Cox, 1978).

The adrenal cortex is regulated largely by the blood level of adrenocorticotrophic hormone (ACTH), which is released by the anterior pituitary, a gland situated at the base of the skull. In recent years it has been recognized that hormone secretion in the anterior pituitary is controlled by the hypothalamus (an area of grey matter in the brain), which secretes chemical agents; these are known as releasing and inhibiting factors, but also as hypophyseotropic hormones (Ganong, 1985). The anterior pituitary releases growth hormone and prolactin (a milk-producing hormone) and four other hormones including ACTH, which control the activities of other glands. They have all been associated with the stress response, and may play a role in resistance, although "it is the anterior pituitary-adrenocortical system which appears to dominate, if not in reality, then certainly in terms of the attention paid to it in research" (Cox, 1978:60). The adrenal medulla, which is closely connected

1 The autonomic nervous sytem activates glands and smooth muscle, and is generally considered to be involuntary although this idea has been challenged (Boddy, 1983). It is divided into sympathetic and parasympathetic systems.

2 The endocrine system consists of the ductless glands which secrete hormones. Examples are the pituitary gland and the thyroid gland.

to the sympathetic nervous system, produces adrenaline and noradrenaline. In humans, about ten times more adrenaline than noradrenaline is secreted by the medulla, but levels of the latter in plasma and urine are greater than the former because of the additional secretion of noradrenaline by the sympathetic nerve endings (Cox *et al*, 1983). Adrenaline increases metabolism, causes the liver to release glycogen as glucose, and dilates the small arteries of the heart and muscles, and the bronchial tubes. The amount of adrenaline secreted varies, and is elevated by fear, anger or excitement. Consequently, the adrenal glands are often referred to as the "glands of flight and fight" (Armstrong and Jackson, 1972:265).

In clinical situations, the activity of the neuro-endocrine system can be measured directly, by assessment of the level of hormonal excretion, for example (Cox, 1978; Scrimshaw *et al*, 1966). The anthropologist has to rely on the indirect evidence of neuro-endocrine activity which can be detected in the teeth and bone; such activity generally causes the release of catabolic hormones (those which release energy), so that growth, an anabolic process (that is, one requiring energy) is inhibited (Goodman *et al*, 1984). Bone growth is a good illustration of this process; bone is "a highly sensitive tissue, adapting itself rapidly to variations in the general well-being and so constituted that in times of special strain it can relinquish its growth activity entirely and relieve the general economy by elimination of that added burden" (Park, 1954:281). One frequently recorded marker of "stress" is dental enamel hypoplasia. This occurs when the synthesis of protein, the basic building-block of the body, is disrupted by elevated levels of hydrocortisone following activity of the adrenal cortex. This in turn causes the reduction or cessation of formation of the enamel protein matrix by ameloblasts, producing a hypoplastic lesion (Rose *et al*, 1985).

Selye's Model of Stress

We can now look in a little more detail at Selye's model of stress, and examine some of the criticisms which have been levelled at it. To recapitulate, Selye proposed stress to be a non-specific response to stimuli, the manifestation of this response forming the General Adaptation Syndrome (Figure 2.1). His notion of stress as a response, rather than as a stimulus, is not shared by more contemporary researchers. Significantly for the biocultural approach, Cox and his colleagues (1983:256) summarised critical views of stress as a response by describing such models as deficient when compared to current "psychological and psychophysiological models of Man" which emphasise the "dynamic nature of the person's interaction with his environment, and...the mediating role of individual psychological processes". Note that it is not the response itself which is controversial, the issue being is the response itself stress, or is it evoked by stress?

The keystone of Selye's model is the notion that the stress response is non-specific, that is, it can be produced by a wide range of stimuli, or stressors:

> "By non-specific we mean effects or responses that are characteristic of many stimuli or agents - for example, the manifestations of the alarm reaction, with the secretion of adrenocorticotrophic hormone (ACTH), corticoids and catecholamines..." (Selye, 1976:53).

In his first publication Selye described stressors as "nocuous" (1936:32), but later reported that pleasurable stimuli "...a game of tennis or even a passionate kiss..." (1957:53) may elicit the stress response. Subsequent studies have also shown that experiences of a generally enjoyable kind are almost as effective as unpleasant experiences in intensifying catecholamine (e.g. adrenaline) excretion (Cox, 1978), but this finding is rarely referred to in the clinical literature and, as far as I am aware, has not been noted in anthropological studies.

The notion of non-specificity has been questioned. For example, it is proposed that the general non-specific pattern of neuroendocrine response to stressors cannot be demonstrated with the facility which Selye proposed, specific patterns of reaction having been shown to be produced by different stressors (Cox *et al*, 1983). Numerous different types of physical stimuli have been shown to increase the activity of the medulla (Cox *et al*, 1983), but against the concept of non-specific catecholamine secretion is the hypothesis that anxiety-producing situations are connected with adrenaline release, while noradrenaline is secreted in circumstances of aggression (Cox, 1978). Initially, Selye focused on physical stressors (such as infection, injury, cold or heat) to the extent that Cox (1987) claimed he ignored the part played by psychological processes, although in a later paper Selye stated that in humans "psychologic stressors, and particularly emotional arousal, are of primary importance. This is especially true of the most common stressors of daily life (frustration, anger, fear and hate)" (1976:56). Mason's (1974) experimental work with monkeys indicated that the pituitary-adrenal cortical axis does not react non-specifically to physical stimuli such as heat and fasting, and that in the case of fasting, for example, it is apparently the situation as a whole, with its unpleasant psychological aspects, which elicits a response, rather than the simple absence of food. Thus Mason proposed that the notion of physiological "non-specificity" which is fundamental to Selye's theory is undermined, for it may be that the non-specific response is behavioural, rather than endocrine. The psychological component raises a further criticism of Selye's hypothesis of non-specificity. Selye himself recognised that his theory was weakened by the failure to recognise the mechanism by which so many different stressors could pass the message to neuroendocrine centres of the need for an adaptive response (Mason, 1971). Selye (1976:54) called this unknown mechanism the "first mediator" of stress-responses which Mason (1971) has suggested may be just the psychological equipment concerned with stimulation or emotional reactions to dangerous or disagreeable circumstances in everyday life. This is not to say that psychological processes are not physiological, but rather it highlights the level within the central nervous system at which integration of neuroendocrine responses occur; the lower level endocrine responses to heat or exercise, for example, are in general more consistent inter-individually than are the higher level responses to distinct "psychosocial stimuli or life events" (Mason, 1975:32). Selye (1975) countered Mason's interpretation of the "first mediator" as psychological on the grounds that stress occurs in lower animals and plants lacking a nervous system, although he offered no operative mechanism; in addition, he pointed to stress produced by anaesthesia, which he proposed is a well-known surgical phenomenon.

It now appears that the majority of hormones, possibly all of them, characteristically respond to multiple stimuli (Mason, 1975:27), and in this sense the response of hormones such as adrenaline, growth hormone and hydro-cortisone could be said to be "non-specific" in that the secretion of each may be stimulated by more than one stimulus. However, Mason has found no evidence "that any single hormone responds to *all* stimuli in *absolutely* non-specific fashion, as implied in Selye's definition of stress as occurring in response 'to any demand'".

The third basic premise of Selye's stress theory is the General Adaptation Syndrome (G.A.S.), which proposes that although the stress response may be triggered by a wide range of stressors, and is thus non-specific, the manifestation of the response is highly specific and forms a recognisable syndrome:

> "Stress is the state manifested by a specific syndrome which consists of all the non-specifically induced changes within a biologic system. Thus stress has its own characteristic form and composition but no particular cause" (Selye, 1957:54).

Selye (ibid.) described this as an "operational definition; it tells what must be done to produce and recognize stress"; it is the manifestations of a state which make it recognizable, so it follows that the manifestations of the stress syndrome, the G.A.S., allow the state of stress to be identified.

The G.A.S. has three distinct phases (Fig.1). The first, the alarm reaction, is chiefly an increase in activity of the sympathetic-adrenal medullary axis, and is accompanied by a reduced level of resistance, which may collapse if the stressor is sufficiently severe, in which case death will ensue. However, if adaptation is possible in the face of continued exposure to the stressor, the second phase, that of resistance, will be entered. The bodily changes of the first stage are replaced by those indicative of adaptation, such as a return to normal body weight (Selye, 1973); this second phase is primarily one of increased anterior pituitary-adrenal cortical activity (Cox *et al*, 1983). Long-term exposure to stress uses up the energy needed for adaptation to it, so that the third phase, that of exhaustion, will occur, leading to death. Only the most severe stress leads to the stage of exhaustion; most stressors produce changes which correspond to the first two stages only, a frequently experienced situation in normal human life (Selye, 1957). Stress exists at any given time during the three stages of the G.A.S. Selye (1957:54) emphasised that stress and the G.A.S. are not synonymous, because "these reactions are characterized by certain measurable organ-changes which are caused by stress and hence could not themselves be stress".

According to Cox and his colleagues (1983), the G.A.S. does not describe precisely the body's response to stress. Referring to Mason's (1968) work, they proposed that what happens in times of stress is increased patterns of catabolic activity, and the reversal of the situation when stress is ameliorated. Mason (1971:330) argued that the G.A.S. is incompatible with the well-established principles of homeostatic regulation: "how could the same hormonal response have adaptive utility in response to diverse stimuli which place diametrically opposite metabolic needs on the body as do, for example, heat and cold?", and he believed the explanation to be a general adaptive response of behavioural or psychological origin. A final but important point to note is the part played by the hypothalamus in the stimulation of hormonal secretion; for example, a failure at this level will cause failure of hormonal secretion by the anterior pituitary (Russell, personal communication). This is significant in terms of the level at which Selye identified the body's response to stressors.

The effect of Selye's controversial theory has been encapsulated by Engel:

> "In his theory of stress ... Professor Selye has presented a concept which is almost breathtaking in its scope. It has elicited extravagant praise and critical condemnation. It has permeated medical thinking and influenced medical research in every land, probably more rapidly and more intensely than any other theory of

disease ever proposed" (Engel, 1956, cited by Mason, 1975:10).

Selye was largely responsible for the surge of interest in physiological stress in the 1950s yet now, a little more than thirty years after Engel's statement, his work is largely overshadowed by subsequent research which has revealed the complexity of the neuroendocrine response (Russell, personal communication), and by a turning away from physiological to psychosocial stress research (Mason, 1975a). The development in the early 1950s of more sensitive and precise methodologies than had been available to Selye confirmed his findings of elevated levels of corticosteroids in blood and plasma in response to a variety of stimuli. As a result, although it was later censured, there was a wide acceptance of the concept of a non-specific response by the pituitary-adrenal cortical system, and a decline in the testing of hypotheses of physiological stress (Mason, 1971).

Anthropological Concepts of "Stress"

The decline in clinical interest in physiological stress has been matched by an enthusiastic uptake on the part of anthropologists. Buikstra and Cook (1980) saw this as a response to criticisms of the value of skeletal studies by the "new" anthropologists. So once again Selye has been influential, for his is the concept of stress which forms the basis, explicitly or implicitly, of many interpretations of stress in the skeletal and dental remains of past populations (e.g. Buikstra and Cook, 1980; Clark, 1985; Clarke, 1978; Condon, 1981; Gilbert, 1985; Goodman *et al*, 1980; Goodman *et al*, 1984; Martin *et al*, 1985; Powell, 1985; Rose and Pasley, 1980; Rose *et al*, 1985). In their survey of American palaeopathology, Buikstra and Cook (1980:444) described the increasing use of indicators of stress, stating "`Stress' in this sense is Selye's concept: any extrinsic variable or combination of variables which causes the organism to react". What they have, in fact, described is Selye's notion of a "stressor", rather than stress, but the point of this reference to their work is that they see Selye's model as central to anthropological interpretations of non-specific skeletal markers.

With the exception of Clark (1985), there has been no published recognition of the criticisms which have been levelled at Selye's model, and to my knowledge, only Clark (1985), Gilbert (1985) and Hillson (this volume) have referred explicitly to psychological stress (defined by Mason (1975a:11) as "the impact of psycho-social influences upon the organism"), anthropologists habitually using "stress" in the sense of a physiological phenomen (Table 2.1): "Stress is a central, stereotypic *physiological* response" (Goodman *et al*, 1980:515, citing Levi, 1972, my emphasis). Hummert and Van Gerven (1983:471) provide a further example of this view of stress; when standards of growth derived from living populations "are used in conjunction with anthropometric and roentgenographic data on the growth response of children suffering from malnutrition, disease and *other physiological stresses*, a fuller understanding of growth and morbidity among prehistoric peoples is facilitated. Such information may provide important insights into the biocultural adaptation of earlier human societies" (my emphasis). Is this an adequate view of stress upon which to build biocultural interpretations of past populations? A review of the clinical literature indicates that it is not.

Psychological Influences on Health

The relationship between health and an adverse physical environment is well-known, and demonstrated, for instance, by the fall in the death rate consequent upon the introduction to Victorian London of adequate sewerage and a clean water supply (Porter, 1987). It has been argued convincingly, by Dubos (1961) and Hinkle (1961) for example, that the *social* environment also has far-reaching consequences for health: "one finds no illness which is not in some degree influenced by the way that men react to what goes on around them, and none that does not occur sometimes in association with manifestations of mood, thought or behaviour (Hinkle, ibid:295). Inherent in such reactions are patterns of adaptation which are regulated by the activity of the endocrine glands, the nervous system and "the gross behaviour of the person" (ibid.:294).

Individuals interact with their social environment at different levels: at an intimate, inter-personal level and more generally with the society or community to which they belong. Psychological stress may be experienced at both levels. Gilbert (1985) proposed that although it is always present in social communities, it is less prevalent in stable societies than in those experiencing sociocultural change. He illustrated his idea by reference to Moss' (1973) model of stress as a result of failure to communicate:

> "...because each individual is enmeshed in some social network(s), at any time the individual fails to interpret correctly the meaning and action of others, his probability of correct response is reduced, consequently increasing his degree of anxiety about responding appropriately. This anxiety is transformed into activation of the sympathetic nervous system with attendant production of adrenaline that over time reduces his resistance to disease. If during times of culture change the individual participates in various networks that are incompatible or fail to provide the proper social clues

for behaviour, his degree of stress will increase" (Gilbert, 1985:341).

Skeletal and dental markers represent insults experienced during the years of growth, and unfavourable familial circumstances may have a profound impact upon the growing child. This is demonstrated clearly by a study of thirteen children with unstable home lives who presented with symptoms suggestive of idiopathic hypopituitarism[3], notably short stature (Powell *et al*, 1967). A reduction in release of ACTH and growth hormone were the most marked deficiencies noted. Removal to a more favourable environment led to an acceleration of growth without the administration of growth hormone. Accordingly it was posited that the "emotional disturbance in these children may have had an adverse effect upon release of pituitary tropic hormone via the central nervous system" (ibid.:1282) which was manifested in growth failure.

Further evidence of the effects of an unfavourable social environment has been presented by Widdowson (1951) in her famous study of orphanage children in post-war Germany. She observed reduced weight gain, despite calorific supplementation, in children subjected to "harsh and unsympathetic handling" (ibid.:1318) by the head of their orphanage; by contrast, eight favoured children gained more than double the standard amount of weight on the supplemented diet. In this study, growth in height was not affected by emotional factors to the same degree as weight, but similar trends are apparent.

The influence of psychological factors on protein metabolism has been demonstrated by Scrimshaw and colleagues (1966). Their investigation of 26 male students before and during final examinations revealed that in 15 cases, despite consistency of diet, this period was associated with metabolic changes, notably elevated nitrogen excretion, indicating the secretion of catabolic hormones. Although this is a study of adults whose growth obviously would not be influenced, it is additional testimony of the power of psychological factors to affect physical processes.

Given that the cause of the disturbances which leave their traces in dental and osseous tissue cannot be categorically identified, how important is it to recognise that psychological factors may play a part in the formation of markers of insult? I believe it to be essential because markers have been interpreted in terms of nutritional status and of subsistence change, particularly with regard to the transition from hunting and gatherering as a mode of subsistence to one of food production (e.g. Cohen and Armelagos, 1984); Widdowson's study shows that a diet may be nutritionally sound but consumed in unfavourable psychological conditions which inhibit growth.

Because markers of insult can form only while growth is ongoing, they reflect metabolic imbalance in childhood; peak frequencies of markers in infants and young children have been attributed to a "weaning crisis" (e.g. Carlson *et al*, 1974; Cook, 1979; Rose, 1977; Rudney, 1983), brought about by the synergistic interaction of such factors as extended lactation by undernourished mothers, an unsuitable or nutritionally inadequate weaning diet and weanling[4] diarrhoea. It may be that psychological factors are a part of this "weaning crisis" and contribute to the formation of markers of insult, especially in a child already physically debilitated. Conflict between parent and child may occur at weaning, either because the infant wishes to continue breast feeding or as a result of sibling rivalry (Smith, personal communication). Enforced weaning may not be the only change which results from the birth of a sibling; studies have shown that behavioural changes are common, especially with regard to the interaction of mother and child (Dunn and Kendrick, 1982), the arrival of a new baby often pressurising a first born into greater independence (Dunn, 1984). The first born is more likely than its younger sibling to be reprimanded when arguments between siblings occur, and there is frequently a marked reduction in the amount of attention paid to a firstborn with the birth of a sibling (*ibid.*). Any of these factors could have an adverse psychological impact, which could disrupt growth and thus influence the formation of markers of insult. A final, intriguing point to note is that the literal translation of "kwashiorkor" (the protein deficiency disease common in developing countries) is "first-second" (FAO/WHO/UNICEF, 1870:27); it is the disease of the toddler superseded, while still being breast fed, by the birth of another baby. In tribal societies mothers often nurse their children for three years or more, so that the child is living in a state of intimate dependency on the mother. Mothers usually wean when they know that they are pregnant, so that the child may adjust, but although relatives take on much of the care of the child at this time, the mother-child bond remains strong; when the sibling is born, however, the mother's whole attention is turned to the newborn (Eibl-Eibesfeldt, 1989).

Of course, psychological factors cannot be extracted from archaeological data in the way that

3 A condition resulting from insufficient pituitary secretion.

4 A "weanling" is a child between the time that a food other than breast milk is consistently taken and the time that it completes the transition to a general diet (Gordon, 1969:220).

changes in subsistence may be, although circumstances which could be expected to induce psycho-social stress in a population (bearing in mind the individuality of response) might be identifiable.

Postscript: The Individuality of Response

The definitions of health which were offered at the beginning of this paper emphasised the dynamic nature of the condition; an individual *rallies* from insult (Audy, 1971) or *adjusts* to his or her environment (Goodman, this volume; Wylie, 1970). Dubos (1965:xvii) has differentiated between an individual's reaction to his or her environment, and the response to it:

> "The components of the body machine react with the environment, in the same way as do similar components in any other living or inaminate system. But living man responds to his environment...human adaptation can seldom be regarded as merely the result of man's body reacting with environmental forces".

This may explain Hinkle's (1961) observation of prolonged good health on the part of individuals who have experienced physical, financial and emotional deprivation, while others succumb to repeated illness despite living in what would seem to be extremely favourable circumstances. Dubos' emphasis on the personality of the individual reiterates the particularity of response, and the essential fact that a universally "stressful" situation does not exist. This is echoed by Hinkle (1961:291). Significantly for biocultural analyses of skeletal remains, he states: "`Environmental stress' and `emotional stress' simply do not exist in the sense that mechanical stress exists in a steel beam". In the final analysis, the reaction of individuals to their social and interpersonal environment depends on their own perception of them.

The individuality of response is a factor of considerable importance when markers of insult are interpreted in terms of the health of a community or population as a whole (see Bush and Zvelebil, this volume), particularly in view of the tendency to refer to "stress" in broad categories as detailed in Table 2.1. It has been proposed, however, that although each human being is unique in terms of experience and attitudes, there are mutual responses in individuals with a common cultural background (Hinkle, 1961). Considering an individual within the framework of their society and their culture means that the reactions of that individual which are products of culture and society can be postulated. Hinkle intended this anthropological evidence to be a tool to aid clinicians' evaluation of their patients. Unfortunately for archaeologists, his recommendation that "a systematic assessment of those aspects of experience that make him unique" (ibid.:292) should also be undertaken is impossible, since they are unknowable to archaeologists.

Conclusions

This chapter has emphasised the need to allow that psychological elements have the potential to influence the formation of markers of insult. It is suggested that the use of Selye's model of stress has led to a failure, by and large, to consider such elements, the consequence of which is that markers of insult may have been interpreted erroneously. This in turn has implications for the wider appreciation of past societies. More prosaically, I would argue that if Selye's model is adhered to, the frequently-recorded condition of porotic hyperostosis should not be referred to as a marker of "stress"; although the lesions do indicate the body's attempt to deal with a morbid condition, in this case that of anaemia, they do not reflect the growth disturbance consequent upon hormonal imbalance which results, for example, in hypoplasia of the dental enamel.

In contrast to Selye's inhibiting, physiological definition of stress, that of Cox and his colleagues (1983:255) seems to accord with the general perception of stress in most anthropological studies:

> "...the *quality of human life* is dependent on the ability to resist disease, and to adapt to the physical and psychological demands of the environment. Failure to adapt can result in impaired physical and psychological well-being, in disease and ultimately in death. These demands, their experience, and effects represent *stress*" (emphasis in the original).

Yet it could be argued, as Hinkle (1961) and Mason (1975b) have done in respect of clinical work, that anthropological studies would benefit from the striking of the word "stress" from the vocabulary, given the problems of definition. Even so, Mason (1975b:35) felt that its usage is not totally devoid of worth, for "perhaps the notion of a generic term which somehow ties together the threatening or taxing demands of the environment on living organisms strikes some deep, responsive chord within us which keeps alive the use of stress terminology in spite of all the confusion it creates"; perhaps this explains its continued use by anthropologists. Moreover, semantic aspects aside, it is the notion of the organism reacting and responding to internal and external agents in order to maintain its inherent balance which forms the basis of the concept of health underlying the biocultural approach. Health cannot be considered in a vacuum because the human organism is receptive to environmental influences, physical, cultural and societal; the socio-cultural system can have a powerful impact upon the individual, producing psychological stress which can be manifested physically.

References

Armstrong, K.F. and Jackson, S.M. (1972). *Anatomy and Physiology for Nurses.* London: Bailliere Tindall.

Audy, J.R. (1971). Measurement and diagnosis of health. In *Environ/Mental* eds. P. Shepard and D. McKinley pp. 140-62. Boston: Houghton Mifflin.

Boddy, J. (1983). Information processing and functional systems in the brain. In *Physiological Correlates of Human Behaviour Vol. 1*, eds. A. Gale and J.A. Edwards, pp.49-77. Orlando: Academic Press.

Buikstra, J.E. and Cook, D.C. (1980). Palaeopathology: an American account. *Ann. Rev. Anthropol. 9*:433-70.

Cannon, W.B. (1914). The interrelations of emotions as suggested by recent physiological researches. *Am. J. Psychol. 25*:256-82.

Cannon, W.B. (1928). The mechanisms of emotional disturbance of bodily functions. *New Engl. J. Med.* 198:877-884.

Cannon, W.B. (1929). *Bodily Changes in Pain, Hunger, Fear and Rage.* Boston: C.T. Branford Co.

Carlson, D.S., Armelagos, G.J. & Van Gerven, D.P. (1974). Factors influencing the etiology of cribra orbitalia in prehistoric Nubia. *J. Hum. Evol. 3*:405-10.

Cassel, J., Patrick, R. & Jenkins, D. (1960). Epidemiological analysis of the health implications of culture change: a conceptual model. *Ann. N.Y. Acad. Sci. 84*:938-49.

Chadwick, J. and Mann, W.N. (1950). *The Medical Works of Hippocrates: a new translation from the original Greek.* Oxford: Blackwell Scientific.

Clark, G.A. (1985). *Heterochrony, Allometry and Canalization in the Human Vertebral Column: Examples from Prehistoric Amerindian Populations.* Ph.D. Dissertation, University of Massachusetts, Amherst. Ann Arbor: University Microfilms.

Clarke, S.K. (1978). *Markers of Metabolic Insult: the Association of Radiopaque Transverse Lines, Enamel Hypoplasias and Enamel Histopathologies in a Prehistoric Human Skeletal Sample.* Ph.D. Thesis. University of Colorado, Boulder. Ann Arbor: University Microfilms.

Cohen, M.N. and Armelagos, G.J. (1984). *eds. Paleopathology at the Origins of Agriculture.* Orlando: Academic.

Condon, K.W. (1981). *Correspondence of Developmental Enamel Defects between the Mandibular Canine and First Premolar.* Unpublished M.A. Dissertation, Department of Anthropology, University of Arkansas, Fayetteville.

Cook, D.C. (1979). Subsistence base and health in prehistoric Illinois Valley: evidence from the human skeleton. *Med. Anthropol. 4*:109-24.

Cook, D.C. (1981). Mortality, age-structure and status in the interpretation of stress indicators in prehistoric skeletons: a dental example from the Lower Illinois Valley. In *The Archaeology of Death.* eds. R. Chapman, I. Kinnes & K. Randsborg. pp.133-44. Cambridge: Cambridge University Press.

Cox, T. (1978). *Stress.* London: MacMillan Press Ltd.

Cox, T., Cox, S. & Thirlaway, M. (1983). The psychological and physiological response to stress. In *Physiological Correlates of Human Behaviour. Vol. 1* eds. A. Gale and J.A. Edwards. pp. 255-76. London: Academic Press.

Dubos, R. (1959). *Mirage of Health.* London: George Allen and Unwin.

Dubos, R. (1965). *Man Adapting.* New Haven and London: Yale University Press.

Dunn, J. (1984). *Sisters and Brothers.* London: Fontana.

Dunn, J. and Kendrick, C. (1982). *Siblings: Love, Envy and Understanding.* London: Grant McIntyre.

Eibl-Eibesfeldt, I. (1989). *Human Ethology.* New York: Aldine de Gruyter.

Engel, F.L. (1956). *A critique of some concepts relating to the etiopathogenesis of the diseases of adaptation.* Proc. Second World Congr. Med. Psychol. Buenos Aires.

FAO/WHO/UNICEF Protein Advisory Group (1970). *Lives in Peril: protein and the child.* Rome: FAO.

Ganong, W.F. (1985). *Review of Medical Physiology.* 12th edition. Los Altos, California: Lange Medical Publications.

Gilbert, R. I. (1985). Stress, paleonutrition and trace elements. In *The Analysis of Prehistoric Diets.* eds. J. Mielke and R. Gilbert. pp. 339-58. Orlando: Academic Press.

Goodman, A.H., Armelagos, G.J. & Rose, J.C. (1980). Enamel hypoplasias as indicators of stress in three pre-historic populations from Illinois. *Hum. Biol. 52*:515-28.

Goodman, A.H., Armelagos, G.J. & Rose, J.C. (1984). The chronological distribution of enamel hypoplasias from prehistoric Dickson Mounds populations. *Am. J. Phys. Anthropol. 65*:259-66.

Gordon, J.E. (1969). Social implications of nutrition and disease. *Arch. Environ. Health 18*:216-34.

Hinkle, L.E. (1961). Ecological observations of the relation of physical illness, mental illness and the social environment. *Psychosomat. Med. 23*:289-97.

Hummert, J.R. and Van Gerven, D.P. (1983). Skeletal growth in a medieval population from Sudanese Nubia. *Am. J. Phys. Anthropol. 60*:471-78.

Huss-Ashmore, R., Goodman, A.H. & Armelagos, G.J. (1982). Nutritional inference from Levi, L. (1972). *Stress and Disease in Response to Psychosocial Stimuli.* Oxford: Pergamon Press.

Levi, L. (1972). *Stress and Disease in Response to Psychosocial Stimuli.* Oxford: Pergamon Press.

Mack, M.E., Blakey, M.L. & Carter, S. (1988). The significance of dental defect width in relation to growth and developmental indicators in the skeleton (abstract). *Am. J. Phys. Anthropol. 75*:243.

Martin, D.L., Goodman, A.H. & Armelagos, G.J. (1985). Skeletal pathologies as indicators of quality and quantity of diet. In *The Analysis of Prehistoric Diets.* eds. J.H. Mielke and R. Gilbert. pp. 227-79. Orlando: Academic Press.

Mason, J.W. (1968). A review of psychoendocrine research on the pituitary-adrenal cortical system. *Psychosom. Med. 30*:576-607.

Mason, J.W. (1971). A re-evaluation of the concept of "non-specificity" in stress theory. *J. Psychiatr. Res. 8*:323-33.

Mason, J.W. (1974). Specificity in the organization of neuroendocrine response profiles. In *Frontiers in Neurology and Neuroscience Research.* eds. P.

Seeman and G. Brown. pp. 68-80. Toronto: University of Toronto.
Mason, J.W. (1975a). A historical view of the stress field. Part I. *J. Hum. Stress. 1*:6-12.
Mason, J.W. (1975b). A historical view of the stress field. Part II. *J. Human Stress.* 22-36.
Mensforth, R.P. (1986). The pathogenesis of periosteal reactions in earlier human groups: diagnosis, epidemiological and demographic considerations (abstract). *Am. J. Phys. Anthropol. 69*:239.
Mensforth, R.P., Lovejoy, C.O., Lallo, J.W. & Armelagos, G.J. (1978). The role of constitutional factors, diet and infectious disease in the etiology of porotic hyperostosis and periosteal reactions in prehistoric infants and children. *Med. Anthropol.* 2:1-59.
Moss, G.E. (1973). *Illness, Immunity and Social Interaction.* New York: John Wiley and Sons.
Palkovich, A.M. (1987). Endemic disease patterns in palaeopathology: porotic hyperostosis. *Am. J. Phys. Anthropol. 74*:527-37.
Park, E.A. (1954). Bone growth in health and disease. *Arch. Dis. Child. 29*:268-81.
Porter, R. (1987). *Disease, Medicine and Society in England 1550-1860.* Basingstoke: Macmillan Education.
Powell, G.F., Brasel, J.A. & Blizzard, R.M. (1967). Emotional deprivation and growth retardation simulating idiopathic hypopituitarism. I. Clinical evaluation of the syndrome. *New. Eng. J. Med. 276*:1271-78.
Powell, G.F., Brasel, J.A. & Blizzard, R.M. (1967). Emotional deprivation and growth retardation simulating idiopathic hypopituitarism. II. Endocrinologic evaluation of the syndrome. *New. Eng. J. Med. 276*:1279-83.
Powell, M.L. (1985). *Health, Disease and Social Organization in the Mississippian Community at Moundville.* Ph.D. dissertation. Northwestern University, Illinois.
Rose, J.C. (1977). Defective enamel histology of prehistoric teeth from Illinois. *Am. J. Phys. Anthropol. 46*:439-66.
Rose, J.C. and Boyd, L. (1978). Dietary reconstruction utilizing histological observation of the enamel and dentin (abstract). *Am. J. Phys. Anthropol. 48*:431.
Rose, J.C. and Pasley, J.N. (1980). *Stress and dental development: an experimental paleopathological model.* Paper presented at 49th Annual Meeting of the American Association of Physical Anthropologists, Niagara Falls, New York.
Rose, J.C., Condon, K.W. & Goodman, A.H. (1985). Diet and dentition: developmental disturbances. In *The Analysis of Prehistoric Diets* eds. J.H. Mielke and R. Gilbert. pp. 281-305. Orlando: Academic Press.
Rudney, J. (1981). *The Paleoepidemiology of Early Childhood Stress in Two Ancient Nubian Populations.* Ph.D. dissertation, University of Colorado, Boulder.
Scrimshaw, N.S., Habicht, J.P., Piche, M.L., Cholakos, B. & Arroyave, G. (1966). Protein metabolism of young men during university examinations. *Am. J. Clin. Nutr. 18*:321-24.
Selye, H. (1950) *Stress.* Montreal: Acta Inc. Med. Publ.
Selye, H. (1957). *The Stress of Life.* London: Longmans, Green and Co.
Selye, H. (1973). The evolution of the stress concept. *Am. Sci. 61*:692-99.
Selye, H. (1976). Forty years of stress research: principal remaining problems and misconceptions. *Can. Med. Assoc. J. 115*:53-6.
Widdowson, E.M. (1951). Mental contentment and physical growth. *Lancet 1*:1316-18.
Wylie, C.M. (1970). The definition and measurement of health and disease. *Pub. Hlth. Rep. 85*:100-04.

Acknowledgements

The doctoral research from which this paper is revised was funded by the Science and Engineering Research Council. I thank Dr. T. Cox, Department of Psychology, University of Nottingham, Prof. R.G.G. Russell, Department of Human Metabolism and Clinical Biochemistry, Royal Hallamshire Hospital, Sheffield and Prof. J.M. Tanner, Institute of Child Health, University of London, for their helpful comments on an earlier version. I am grateful to Dr. Peter Smith, Reader in Psychology, University of Sheffield, for useful discussion.

Table 2.1. The anthropological usage of "stress".

Usage	Publication
As a non-specific physiological state:	Condon, 1981 Goodman *et al*, 1980 Goodman *et al*, 1984 Huss-Ashmore *et al*, 1982 Rose and Boyd, 1979 Rose and Pasley, 1980 Rose *et al*, 1985
"malnutrition, disease and other physiological stresses"	Hummert and Van Gerven,1983:471
"biological stress"	Hummert and Van Gerven, 1982:1
"systemic stress from specific or non-specific infection and/or from general or specific nutritional deficiencies"	Powell, 1985:203
"...response to nutritional stress"	Mensforth *et al*, 1982:1
"chronic or endemic dietary stress"	Palkovich, 1987:527.
"infectious disease stress"	Mensforth, 1986:239
"childhood stress and morbidity"	Hummert and Van Gerven, 1983:471
"the effects of stress that dental enamel defects represent"	Mack *et al*, 1988:243
"a human population under stress"	Mensforth *et al*, 1982:2
"high status individuals experienced less stress"	Cook, 1981:140
"psychosocial stress"	Gilbert, 1985:341

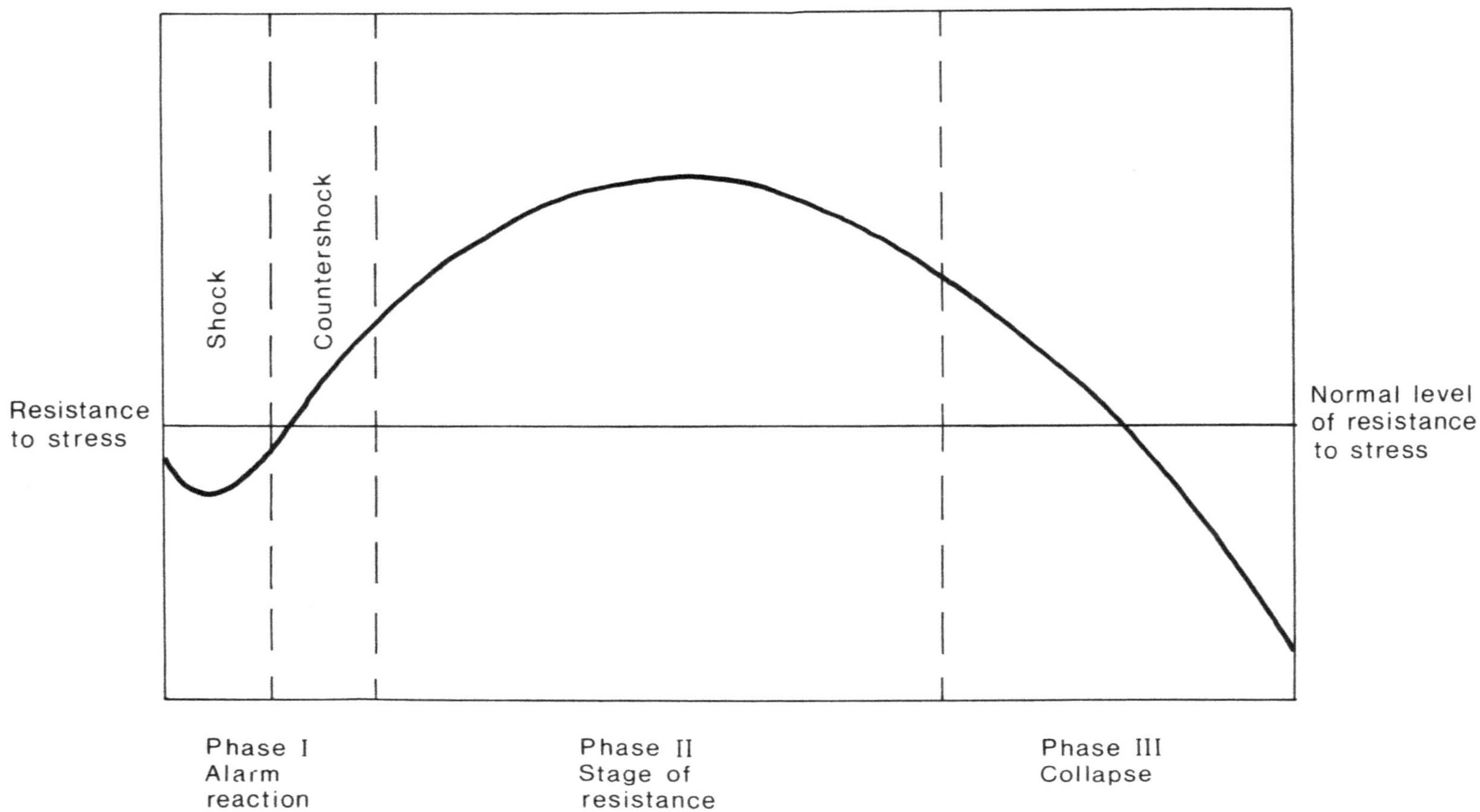

Figure 2.1 The General Adaptation Syndrome (from Selye, 1957).

3. ACCOUNTING FOR THE EVOLUTION OF PSYCHO-SOMATIC PHENOMENA: DID UNCO-OPERATIVE UPPER PALAEOLITHIC PEOPLE BECOME ILL AND DIE?

Michael E. Hyland and Win Scutt.

The relationship between archaeology and psychology has received little attention from either archaeologists or psychologists. The theoretical potential of research at the interface of these two disciplines is often obscured, at least in part, due to a simplified view members of either discipline have of the other discipline.

To a psychologist, behaviour is the consequence of an interaction between genetic and environmental factors. The way genetic and environmental factors interact to produce behaviour varies along the phylogenetic scale. In the case of simple creatures, such as spiders or frogs, behavioural patterns are genetically determined (and referred to as "fixed action patterns"). In the case of humans, where there are few fixed action patterns, the genetic contribution is primarily one of a *capacity*, and behaviour is the consequence of the interaction of that capacity with environmental conditions (Bateson, 1979). For example, humans, but not dogs, have the genetically transmitted capacity to learn language given the right environmental conditions.

For many psychologists, the genetic contribution to behaviour is assumed to have evolved because it was biologically adaptive, that is, advantageous to the gene pool. But contrary to ecological theory, the concept of biological adaptation is divorced from the ecological environment. For example, Lorenz (1966; see also Bailey, 1987) suggests that aggression has a genetic basis because it is biologically adaptive; on the other hand, Rushton, Russell and Wells (1984) suggest that altruism has a genetic basis because it is biologically adaptive. Both psychologists fail to recognise that what is biologically adaptive in one ecological environment may not be biologically adaptive in another. In short, the evolution of the genetic contribution to behaviour cannot be understood divorced from its environmental context. It can be understood only in terms of a series of evolutionary environments as presented by the archaeological record. To be consistent with evolutionary theory, the psychologists's concept of the biological utility of behaviour will have to depend on archaeology.

To most archaeologists, an organism's behaviour is assumed to be dictated by its physical form and its cultural or "extra-somatic" (White, 1959) means of adaptation. Thus, in the case of humans, psychological changes (e.g., improvement in thinking skills) are assumed to follow changes in morphology (e.g. increase of skull capacity), or some form of cultural change which enhances thinking skills (e.g. early infant stimulation). Thinking skills are thought not to improve except with morphological and cultural change. A related assumption is that similar morphologies are indicative of similar genotypes.

The failure to recognise that psychological capacities may have a genetic component is a difference between archaeology and psychology. Certainly it is theoretically possible that similar morphologies may be accompanied by different genetically determined psychologies. However, an example of the weakness of relying on morphological similarity to infer total genetic similarity is provided by the brown trout and sea trout (the latter is described by anglers as a "sea run brown trout") which have the same morphology and so are both classified as *Salmo trutta* but have different behaviour patterns. Evidence for genetic dissimilarity between sea and brown trout is provided by smolt transformation which occurs in sea trout even when raised in freshwater hatcheries but never in brown trout. It is now thought that the brown trout evolved from the sea trout as part of the recolonization of rivers after the last ice age (Harris and Morgan, 1989).

The success of a species depends not only on its morphology but also on its behaviour, and so psychological evolution must be relevant to an interpretation of the archaeological record. Thus for instance, the early *Homo sapiens sapiens* from Qafzeh and Skhul who have now been dated to about 90,000 BP and whose lithic artifacts represent a Levallois-Mousterian (Middle Palaeolithic) industry (Stringer, 1988; Stringer, Grun, Schwarcz & Goldberg, 1989) may not necessarily represent the same gene pool as anatomically modern humans dating from the Upper Palaeolithic. Anatomical similarity is a necessary but not a sufficient condition for genetic similarity, and a change in behaviour with morphological constancy may not be exclusively the result of extra-somatic adaptation but may reflect some intra-somatic adaptation in the form of a change to the genetic substrate of behavioural capacities.

The purpose of this paper is to explore one particular interface between psychology and archaeology. In recent years, there has been an accumulation of evidence that psychological states have a causal relationship with morbidity (illness) and death (mortality). Following early evidence that bereavement increases mortality in the remaining spouse, there now seems to be considerable (though by no means uncontroversial) evidence that psychological factors play a role in the incidence and recovery of *all* physical disease - such as respiratory tract infections, cold-sores, cancer, heart disease, arthritis and so on. Contrary to the assumptions of traditional physical medicine (referred to in medicine as "the medical model") - a model reflected in some of the other papers in this volume - the incidence of physical illness is not the result of only diet and other physical conditions. The incidence of physical illness is affected by many causes, one of which is the psychological state of the individual. It is now recognised (Baker, 1987) that the term psychosomatic illness should apply not to a small number of illnesses but can be characteristic of all physical illness.

If there is a relationship between psychological states and physical illness, this relationship has either arisen by chance or it has evolved. The purpose of this paper is to demonstrate that the relationship between psychological states and morbidity, henceforth referred to as *psychosomatic phenomena,* did not evolve by chance; it evolved through normal evolutionary processes and the period of evolution can be understood from a combination of existing archaeological and psychological data.

Two Ways in Which Presenescent Mortality is Biologically Adaptive

In ours and many other cultures, death is viewed somewhat negatively. This negative approach to death comes about from our own individualistic perspective: people try to preserve their own lives and those of their loved ones. At the level of the individual, the death of a close friend or relative is always a tragedy. There is, however, an alternative and much more positive view of death which comes from a genetic perspective. In the case of sexual reproduction, maintenance of the genetic material (the purpose of the gene) is achieved through the creation of new organisms which are slightly different from their parents, and *from the death of those parents.* From a genetic perspective, death, and in particular, death from senescence, is an essential part of biological adaptation, where biological adaptation is defined as a characteristic which enhances the continuation of the gene pool.

Not only is senescent death biologically adaptive, but in some cases the death of parents is exquisitely timed to improve biological adaptation of the species. The sockeye salmon spawns and dies within a couple of days though a process of accelerated senescence (Cousteau, 1975). By dying, the parent sockeye salmon do not compete for food with their offspring, and the parent's rotting carcase provides a rich source of food. What is significant about the death of the sockeye salmon is that senescence is controlled not by an "internal clock" or some other internal, physiological mechanism but through a behavioural trigger. The behaviour of spawning triggers off an *internal* self-destruct process which takes the form of accelerated senescence.

It has long been recognised that presenescent death can be biologically adaptive. The classic example of this phenomenon is the case of worker bees sacrificing themselves to save a hive under attack. The cause of death when worker bees sacrifice themselves is *external* to the bee (i.e., it is the attacker). However, it is also possible for presenescent death to be biologically adaptive where there is an *internal* mechanism for death, as in the case of post-reproductive sockeye salmon. There are two ways in which internally driven, presenescent death can be biologically adaptive, these two ways occurring in either a *K*- or *r*-environment (Pianka, 1988).

A *K*-environment is characterised by high intraspecific competition for scarce resources, such that population growth is constrained by population size. Under such conditions, it may be biologically adaptive if weaker members of the species become ill and die, as by dying they do not compete with the superior members of the species for scarce resources. Endocrinological changes in socially inferior members of a species is a recognised form of population control (Christian, 1980). A good example of this phenomenon is provided by trout who are social animals and have dominance hierarchies. Trout at the bottom of the hierarchy tend to be immunosuppressed (i.e., have less active immune systems) and are therefore more likely to succumb to infections (Peters *et al*, 1988).

An *r*-environment is characterised by relatively low intraspecific competition for scarce resources such that population growth is unconstrained by population size. If there were a form of *r*-environment where the survival of the species depends crucially on co-operation between group members, then, under such conditions, unco-operative members can threaten the survival of their group and thereby threaten survival of similar genetic material. It would therefore be biologically adaptive if such unco-operative individuals became ill and died as to do so would enhance the survival chances of others and of the gene pool.

Evolutionary Forces in Food Collecting and Food Producing Societies

For *Homo sapiens*, the Upper Palaeolithic was an *r*-environment characterised by dispersal into new regions (Gamble, 1986; Geist, 1978). Glaciation, high winds, and reduced rain-fall during that period would have introduced environmental stress which could have been alleviated to some degree through co-operation. There is considerable evidence that Upper Palaeolithic people (*Homo sapiens sapiens*) and to a lesser extent Neanderthals (*Homo sapiens neanderthalensis*) were highly co-operative in their gatherer/hunter/fisher or *food collecting* economy (Gamble, 1986; Geist, 1978). Both lived in social groups, both engaged in co-operative hunting, both cared for the sick and both used symbolic burial of the dead. There is circumstantial evidence that co-operative skills were particularly developed in the case of Upper Palaeolithic people: camp sites were larger than in the case of the Neanderthals; homicide was less frequent in Neolithic groups; and the wide distribution of artifacts suggests that cross-regional co-operation took place (Geist, 1978). Extrapolations from modern day gatherer/hunter/fishers suggest that *Homo sapiens* survived in a Palaeolithic environment through a combination of intragroup (local) and intergroup (regional) co-operation. For instance, Binford (1983) shows that if there is a catastrophic decrease in food resources in an area occupied by one Inuit group, that group will move as guests into another group's area. Such co-operation is a useful strategy for survival of a species when survival is so precarious. From an evolutionary perspective, the Palaeolithic, and particularly periods of glaciation during the Palaeolithic, was an environment with a strong evolutionary force for co-operation within the species *Homo sapiens* (Binford, 1983; Gamble, 1986).

The more benign climate which marks the onset of the Mesolithic initiated a gradual shift to a Neolithic, primarily *food producing* economy based on agriculture and animal husbandry. The change from food collecting to food producing had a profound effect on population density. Neolithic communities characteristically have an increased population density and aggregations of greater size than either Palaeolithic or Mesolithic communities.

The reason why Neolithic communities developed relatively simultaneously around the world during the present interglacial rather than during an earlier interglacial has been the subject of much controversy. To those other explanations (such as packing, e.g. Binford, 1983; see also Bender, 1975) we now add a psychological explanation. The development of a Neolithic community would have required the psychological capacity for advanced co-operation. That is, it would be necessary for people to have been *skilled* in co-operation. If the psychological capacity for co-operation was an Upper Palaeolithic evolutionary development (i.e., a change in psychology despite a constant morphology) then this could be a contributory factor to the emergence of a food producing economy shortly after the onset of the present interglacial.

However, although Neolithic society depended partly on co-operation, co-operativeness no longer had the biological utility it had had in the Palaeolithic, and the reason for this change relates to group size. If a person was unco-operative in a Palaeolithic group, the relatively small size of the group would mean that the unco-operative person could affect the working of the group as a whole, and hence the group's survival. Survival was so precarious in the Upper Palaeolithic that unco-operativeness would have affected the gene pool. No so in the Neolithic, where the increased population size would have acted as a buffer against an individual's unsocial acts. Thus although the development of a Neolithic community requires co-operation, once that community has developed, co-operativeness has less biological utility.

Moreover, the increased population density in food producing and particularly urban societies would have shifted the environment into a *K*-environment. As population density increased, within-species competition for scarce resources would have enhanced the biological utility of competitiveness. Evidence shows the change from an *r*-environment to a *K*-environment to be a gradual one (Bender, 1975). To take the later Prehistory of Europe as an example, competitiveness can be seen in the shift from generalised to negative reciprocity (Sahlins, 1968; Service, 1966), the growth of exclusive territoriality, numeracy and wealth (Renfrew, 1972) and in the increasing incidence of artifacts reflecting warfare, from weapons in homicide contexts to fortifications (Newell, 1984; Meiklejohn *et al*, 1984). The biological utility of competition is evident in recent times where settlers in North America and Australia competed so successfully with the indigenous populations that they have prospered at the expense of the indigenous populations. Thus, from an evolutionary perspective, food producing and urban societies are characterised by a strong force towards intraspecific competition in the species *Homo sapiens* which frequently takes the form of regional conflict.

When Did the Psychosomatic Phenomena Evolve?

There are two periods in which psychosomatic phenomena could have evolved: *either* in the food collecting environment of the Palaeolithic where co-operation was biologically adaptive *or* in the food producing/urban environment which has gradually developed since the Neolithic and where regional competition has been biologically adaptive. In either case, the psychological states which would be indicative of biological utility, and so evolve to promote or harm physical health, would be rather difficult. If psychosomatic phenomena evolved during the Palaeolithic, then psychological states associated with co-operation, such as feelings of social support, would be healthy and states associated with competition, such as aggression towards others would be unhealthy. By contrast, if psychosomatic phenomena evolved in food producing and urban societies, then feelings of aggression towards others and other manifestations of competitive success would be healthy.

It is merely necessary to examine the psychological literature on the relationship between psychological states and morbidity to see whether the data fits better the Palaeolithic or food producing/urban model. To simplify the somewhat controversial results, unhealthy psychological states include depression (Jemmott & Locke, 1984), particularly states associated with bereavement (Jones, 1987), loneliness (Lynch, 1977), anxiety (Friedman and Booth-Kewley, 1987), hostility and anger (Dembroski and Costa, 1987; Friedman and Booth-Kewley, 1987), inability to express emotion including inability to express hostility (Shaffer *et al*, 1987), suspiciousness (Barefoot *et al*, 1987), and time pressure (Wright, 1988). Healthy psychological states include social support (Seeman and Syme, 1987), optimism (Scheier and Carver, 1987), affiliation motivation (Jemmott, 1987) and effective coping styles including hardiness, commitment and control (Hull *et al*, 1987).

It would appear that although some of the healthy/unhealthy psychological states such as depression and optimism, fit equally psychosomatic evolution during either the Palaeolithic or the later food producing/urban societies, other psychological states are consistent only with evolution during the Palaeolithic. Consider, for example, the characteristic of hostility. Hostility has been presented as being biologically adaptive (Lorenz, 1966) on the basis that it contributes to individualistic selection, that is, selection where only the most competitive members of a species survive to pass on their genes. In food producing and urban environments where there is competition for scarce resources, the hostile competitive individual should be well adapted - as illustrated by the success of "robber barons". Thus, if psychosomatic phenomena evolved during food producing or urban environments, hostility should be an indicator of biological success and should be healthy - very much in the same way that dominance is healthy in the rainbow trout (Peters *et al*, 1988). That hostility is in fact unhealthy in humans, and that hostility would have been maladaptive in a Palaeolithic environment as it would have reduced co-operation, leads to the inference that psychosomatic phenomena evolved during the Palaeolithic. Similarly, only Palaeolithic evolution is consistent with the finding that social support and affiliation motivation are healthy.

Thus the overall pattern of the way psychological states relate to ill health is indicative of the evolution of psychosomatic phenomena during the Palaeolithic but not at some later period. This is because *all* these psychological states *could* have evolved during the Palaeolithic but only *some* could have evolved during later times. The conclusion to be drawn from the psychosomatic data is as follows. People became ill during the the Palaeolithic when they experienced certain sorts of psychological states on a regular basis. Psychosomatic induced sickness during the Palaeolithic would have been an adaptive mechanism for ensuring co-operation in a society where co-operation was essential for survival. The adaptive advantage of psychosomatic illness disappeared gradually with the development of food producing and urban environments, though the genetic capacity to respond psychosomatically has not been lost during the last 8,000 years (a relatively short time period of human evolution). People still react psychosomatically as if they were in a Palaeolithic environment. Psychosomatic illness is like an appendix: a structure which was once biologically adaptive but now acts simply as a health problem.

The Upper Palaeolithic and the Evolution of Longevity

The argument so far has been that psychosomatic phenomena evolved during the Palaeolithic due to the strong evolutionary pressure towards co-operative behaviour in food collecting societies, particularly under conditions of environmental stress. In this section we suggest that the greatest pressure towards co-operation would have occurred amongst Upper Palaeolithic people and therefore psychosomatic phenomena are most likely to have evolved during the Upper Palaeolithic.

The maximum life expectancy of an organism is genetically controlled. (The average life expectancy is, of course, controlled in addition by other factors such as the presence of predators). In most species the onset of senescence is linked to a reduction in reproductive value (i.e., the probability of producing offspring in the following year). The genetic mechanism underlying this relationship is that genes can be "switched on" on the basis of an internal clock, i.e., the gene has an impact on the phenotype only when it is switched on. If pathological genes are switched on only after the organism has achieved its full reproductive capacity, then those pathological genes are never selected out from the gene pool. Thus, old age is referred to as a "genetic dustbin" because of the accumulation of pathological genes which are expressed at that time (Pianka, 1988).

The longevity of anatomically modern humans is a biological anomaly. It is not that the *average* life expectancy is too long (the average life expectancy is, of course, affected by the quality of medical care) but that the *maximum* life expectancy (which is about 100) occurs when the reproductive value of women has been zero for about 60 years and the reproductive value of men has been low for a long period of time. The normal explanation (e.g. Binford, 1983) for the evolution of longevity in anatomically modern humans is that elderly people played some sort of social role in Palaeolithic society, as leaders or mediators, as educators of the young or as repositories of social and environmental information.

Evidence suggests that the evolution of longevity is a recent and relatively abrupt development of *Homo sapiens*. Using analysis of bone chemistry to assess the age of death in Neanderthals, Trinkaus and Thompson (1987) have shown that the oldest Neanderthals lived to at most the early forties - a maximum life expectancy which, interestingly, is similar to that of the chimpanzee (Burton, 1962). By contrast, some anatomically modern humans contemporaneous with those Neanderthals are estimated to have lived well into their sixth decade. Thus, the development of longevity seems to be relatively recent and a characteristic only of *Homo sapiens sapiens*.

It is unlikely that such an anomalous development in longevity could have occurred without good reason. Although it seems probable that there were few elderly people in the Upper Palaeolithic, there were *some* and that small number of elderly (i.e. post-reproductive) people *must* have had sufficient biological utility to counter the normal senescence-reproductive value relationship in evolution. As physical skills decrease after forty, the idea of a social function of elderly people seems most likely, though the precise function such individuals played is just speculation. What we can be reasonably certain about, however, is that elderly Upper Palaeolithic people contributed in some way to the survival of the group, and that their contribution was essential. After all, despite being physically stronger, and despite having a similarly sized brain, the Neanderthals did not survive - but then they did not have elderly people.

Whatever social role elderly people had in Upper Palaeolithic groups, that role could have been useful only if there was co-operation. Social roles such as leadership, mediation, education or knowledge repositories can be fulfilled only if other members of a group co-operate to allow the elderly person to carry out that role. The introduction of elderly people in the Upper Palaeolithic would therefore have introduced an additional strain towards co-operation so it would have been additionally advantageous if un-cooperative individuals became ill and died. Moreover, given the possible roles identified for elderly people, it is clear that few elderly people would actually be needed, and that excess elderly people would have been a drain on resources. Thus, it would also be biologically adaptive if elderly people could be selected in terms of their contribution to the group so that good contributors survive and others do not.

In summary, during human evolution, the Upper Palaeolithic is the period when psychosomatic phenomena would have been most useful as promoters of biological utility. The most likely time for the evolution of the relationship between psychological states and morbidity therefore seems to be the Upper Palaeolithic. After the Palaeolithic and Mesolithic, the biological advantage of the particular psychological state-morbidity relationship which we observe has decreased and after the Neolithic a different form of relationship may have been adaptive. It is certainly possible that the psychological state-morbidity relationship we observe in humans today is a vestigial and weakened form, and that this relationship was stronger in the Upper Palaeolithic. It is also possible that psychosomatic phenomena are stronger in modern day food collecting groups compared to groups with a food producing history.

Summary and Postscript

The functional theory of illness suggests that the relationship between psychological states and or *psychosomatic phenomena* evolved during the Upper Palaeolithic at which time presenescent illness and death were biologically adaptive. Psychosomatic phenomena were biologically adaptive (i.e. enhanced the survival of the gene pool) by protecting the social cohesiveness of groups operating in a harsh environment. Evidence for this theory comes from the particular pattern of psychological states which relate to illness, from inferences about the evolutionary forces affecting human evolution during and since the Palaeolithic, and inferences about the evolution of longevity.

The functional theory of illness differs from sociobiological theories where *traits* (such as co-operativeness or competitiveness) are assumed to be genetically determined and biologically advantageous. We make no assumptions about the extent to which traits are genetically or environmentally controlled. Interestingly, if such traits *are* genetically controlled, then functional psychosomatic phenomena would have a relatively small effect on such traits in the gene pool as selection of people may occur once they have become post-reproductive.

However, the functional theory of illness provides an ideal adaptive mechanism if co-operation and hostility are situationally controlled (i.e., not genetically controlled, see Leakey and Lewin, 1977). Successful adaptation in an Upper Palaeolithic life style would have involved hostility towards other animals but co-operation towards people. Thus, successful adaptation requires some *situation specific* hostility, that is, hostility which occurs only in appropriate situations. A person who was genetically predisposed to be always non-hostile or always hostile would be biologically unadapted to the environment. The functional theory of illness would provide a mechanism for eliminating those biologically unadapted individuals who were inappropriately hostile, but without relying on a change to the gene pool.

The functional theory of illness presents a view where presenescent death and illness is biologically adaptive. To some people this view is counter-intuitive and this is so because it is common to take a phenomenological and individualistic perspective of death. Evidence from the study of other animals shows, however, that presenescent death can be biologically adaptive.

Those who perceive the functional theory of illness to be a fanciful and implausible story should consider the alternative. Either the psychological state-morbidity relationship evolved in response to evolutionary pressures or it evolved by chance. To suggest that this relationship evolved by chance goes against the very principle of scientific investigation. To say that something occurs "by chance" is scientific nihilism. Things which are thought to occur by chance are often shown at a later date to obey normal scientific laws. The "it occurred by chance" explanation is simply not an explanation for the evolution of psychosomatic phenomena. The functional theory of illness may not be correct; but it is the only theory of the evolution of psychosomatic phenomena which exists today.

References

Bailey, K. (1987). *Human Paleopsychology: Applications to Aggression and Pathological Processes.* San Fransisco: Erlbaum.

Baker, G.H.B. (1987). Psychological factors and immunity. *J. Psychosom. Res. 31*:1-10.

Barefoot, J.C., Siegler, I.C., Nowlin, J.B., Peterson, B.L., Haney, T.L., & Williams Jr., J.B. (1987). Suspiciousness, health and mortality; a follow-up study of 500 older adults. *Psychosom. Med. 49*:450-57.

Bateson, G. (1979). *Mind and Nature.* New York: Dutton.

Bender, B. (1975). *Farming in History: From Hunter/Gatherer to Food Producer.* London: John Baker.

Binford, L.R. (1983). *In Pursuit of the Past.* London: Thames and Hudson.

Burton, M. (1962). *University Dictionary of Mammals of the World.* New York: Cromwell.

Christian, J.J. (1980). Endocrine factors in population regulation. In *Biosocial Mechanisms of Population Regulation.* eds. M.N. Cohen, R.S. Malpas & H.G. Klein. pp.55-115. Yale: Yale University Press.

Cousteau, J.Y. (1975). *The Action of Life.* New York: Plenum.

Dembroski, T.M. and Costa, P.T. (1987). Coronary prone behavior: components of the Type A pattern and hostility. *J. Pers. 55*:211-35.

Friedman, H.S. and Booth-Kewley, S. (1987). The "Disease-Prone Personality": a meta-analytic view of the construct. *Am. Psychol. 42*:539-55.

Gamble, C. (1986). *The Palaeolithic Settlement of Europe.* Cambridge: Cambridge University Press.

Geist, V. (1978). *Life Strategies, Human Evolution, Environmental Design.* New York: Springer-Verlag.

Harris, G. and Morgan, M. (1989). *Successful Sea Trout Angling.* London: Blandford.

Hull, J.G., Van Treuren, R.R. & Virnelli, S. (1987). Hardiness and health: a critique and alternative approach. *J. Pers. Soc. Psychol. 53*:518-30.

Jemmott III, J.B. (1987). Social motives and susceptibility to disease: stalking individual differences in health risks. *J. Pers. 55*:267-98.

Jemmott III, J.B. and Locke, S.E. (1984). Psychosocial factors, immunologic mediation and human susceptibility to infectious diseases: how much do we know? *Psychol. Bull. 95*:78-108.

Jones, D.R. (1987). Heart disease mortality following widowhood: some results from the OPCS longitudinal study. *J. Psychosom. Res. 31*:325-33.

Leakey, R.E. and Lewin, R. (1977). *Origins.* London: Macdonald and Jane's.

Lorenz, K. (1966). *On Aggression.* London: Methuen.

Lynch, J.J. (1977). *The Broken Heart: Medical Consequences of Loneliness.* New York: Basic Books.

Meiklejohn, C., Schentag, C.T., Venema, A. & Key, P. (1984). Socioeconomic change and patterns of pathology and variation in the Mesolithic and Neolithic of Western Europe: some suggestions. In *Paleopathology at the Origins of Agriculture*, eds. M.N. Cohen and G.J. Armelagos. pp. 75-100. Orlando: Academic Press.

Newell, R.R. (1984). On the mesolithic contribution to the social evolution of Western European Society. In *European Social Evolution: Archaeological Perspectives*, ed. J. Bintliff. pp. 69-82. Bradford: University of Bradford Press.

Peters, G., Faisal, M., Lang, T. & Ahmed, I. (1988). Stress caused by social interaction and its effect on susceptibility to *Aeromonas hydrophila* infection in rainbow trout *Salmo Gairdneri. Dis. Aquat. Org. 4*:83-9.

Pianka, E.R. (1988). *Evolutionary Ecology* (3rd ed.) New York: Harper and Row.

Renfrew, A.C. (1972). *The Emergence of Civilization: the Cyclades and the Aegean in the Third Millenium BC.* London: Methuen.

Rushton, J.P., Russell, R.J.H. & Wells, P.A. (1984). Genetic similarity theory: beyond kin selection. *Behav. Genet. 14*:179-93.

Sahlins, M.D. (1968). *Tribesmen.* London: Prentice Hall.

Scheier, M.F. and Carver, C.S. (1987). Dispositional optimism and physical well-being: the influence of generalized outcome expectancies on health. *J. Pers. 55*:169-210.

Seeman, T.E. and Syme, S.L. (1987). Social networks and coronary artery disease: a comparison of the structure and function of social relations as predictors of disease. *Psychosom. Med. 49*:341-54.

Service, E.R. (1966). *The Hunters.* London: Prentice Hall.

Shaffer, J.W., Graves, P.L., Swank, R.T. & Pearson, T.A. (1987). Clustering of personality traits in youth and the subsequent development of cancer among physicians. *J. Behav. Med. 10*:441-47.

Stringer, C.B. (1988). The dates of Eden. *Nature 331*:565-66.

Stringer, C.B., Grun, R., Schwarcz, H.P. & Goldberg, P. (1989). ESR dates for the hominid burial site of Es Skhul in Israel. *Nature 338*:576-78.

Trinkaus, E. and Thompson, D.D. (1987). Femoral diaphyseal histomorphometric age determinations for the Shanidar 3, 4, 5 and 6 Neandertals and Neandertal longevity. *Am. J. Phys. Anthropol. 72*:123-29.

White, L.A. (1959). *The Evolution of Culture*. New York: McGraw-Hill.

Wright, L. (1988). The Type A behavior pattern and coronary artery disease. *Am. Psychol. 43*:2-14.

4. HEALTH, ADAPTATION, AND MALADAPTATION IN PAST SOCIETIES

Alan H. Goodman

Introduction

The analysis of health in past societies has too frequently been undertaken in a vacuum. Many paleopathological studies have been undertaken with little concern for either the cause or effect of health problems. Too infrequently have efforts been made to attempt to understand the relationship between health and adaptation, and equally infrequent are analyses of the causes of ill health that focus beyond the immediate environment. Paleopathologists are guilty of "quitting early" (Harris, 1988), before the implications of signs of health and illness have been fully understood.

Paleopathological analyses, nevertheless, have great potential for elucidating past adaptive struggles, and their links to the present human condition. Individual measures of health derived from skeletal remains, ranging from mild growth retardation to mortality, can be viewed as signs of the struggle to adapt. The pattern of individual indications of health and disease provides a basis from which one can infer the adaptive and functional costs of illness in areas such as reproduction, resistance to disease, sociability, work capacity, and learning. The pattern of disease in groups can be used to reconstruct differential access to resources and exposure to stress.

In the following paper I wish to provide some sense of the potential of palaeopathology, and some sense of the challenge of not quitting early. I begin by introducing a model for studying health and disease in relationship to adaptation in past societies. After the model is briefly explained, the functional significance of growth stunting, one measure of health status, is explored. This discussion is based on studies of the functional consequences of short stature in contemporary populations and is meant to illustrate the type of inferences that can be derived from measurements of bones and teeth. I end by suggesting directions for future research and by placing the study of health and adaptation in the larger theoretical debate over the validity of processual approaches in archaeology.

A Model of Health, Nutrition, and Adaptation in Past Scocieties

This adaptivist model was originally presented in 1982 (Huss-Ashmore *et al.*, 1982) and has been recently revised (Goodman and Armelagos, 1989) (Figure 4.2). The purpose of this model is not to codify links between component in the adaptive process, or to suggest that all adaptive dynamics can be traced through the model. Rather, the purpose of the model is to provide a mechanism for systematically examining and reflecting on the adaptive process and the centrality of health to this process.

The model starts with a consideration of "environmental constraints" to adaptation (Figure 4.2). These are typically divided between limiting resources and stressors. The most important limiting resources are likely to be basic ones: food, water, shelter. Stressors may be related to climatic extremes, such as excesses of heat and cold, or low partial pressure of oxygen. All of these climatic and physical stressors are important because they may have an effect on resistance to disease. Other environmental stressors, such as parasites and predators, have more obvious and direct affects on health and longevity.

There are a number of different ways that environmentally initiated constraints and stressor can be adapted to. Culture, through its technological, social and ideological systems, mediates the process of extracting resources from the environment. Cultural systems are generally effective in buffering environmental constraints and stressors, or removing the individuals in a society from contact with the stressor.

There are numerous examples of effective cultural buffering of stressors and limiting resources. Shelter and clothing are generally well suited as buffers against extremes of cold and heat. An interesting example of how shelter can aid adaptation is presented by May (1960) who relates how peasants in the hilly region of North Vietnam have customarily built their houses on stilts, about eight to ten feet above the ground. This is just above the flight ceiling of Anopheles minimus, a very fierce malaria vector. The adaptive significance of building houses on stilts became even clearer when overpopulation developed in the more fertile delta regions and groups from the delta began to migrate to the hills. The delta peoples were accustomed to building their houses on the ground. Subsequently, they found themselves to be within range of the malaria vector and they suffered frequently from malaria. Consequently, the delta peoples have been discouraged from relocating and attribute their health problems to the evil spirits in the hills (May, 1960).

As a general example, the adoption of agriculture is an economic adjustment that embodies a number of changes in the ability of a culture to buffer stresses. The energetic efficiency of agriculture and the amount of food produced per unit area is greater than that of hunting and gathering. Thus, on the face of it, agriculture can provide a buffer against malnutrition, perhaps the most persistent and significant stressor faced by paleolithic hunter-gatherers (McKeown, 1988). However, one nearly invariable demographic effect of agricultural intensification is greater population density. Unfortunately, the increased population density and other ecological and demographic changes that are often secondary to agricultural intensification had unexpected consequences for health, such as improving conditions for the spread of infectious diseases. Furthermore, in West Africa it is likely that clearing of forest for agricultural production brought humans into close contact with mosquitoes, thus leading to the spread of malaria in that area (Livingstone, 1958).

The important point in the above is that the dynamic cultural system, which I have just briefly noted, operates in two ways: as a buffering system and as a producer of new stressors and constraints. As cultures change the nature of environmental adaptation changes. Improved adjustment to one environmental condition is frequently only made by increasing exposure to another condition. Compromised adaptations are frequent; they are the rule. As well, cultures and their adaptations vary in how well different segments of the population are buffered from stressors (Goodman and Armelagos, 1989).

The environmental constraints that are not well buffered by the cultural system, along with the newly produced cultural stressors, reach and contact the individuals in the society. From this point on, the degree to which these constraints are adapted to depends on the individual's level of host resistance. Resistance can be related to one's genetic, developmental, and physiological status. If a stressor is persistent and life threatening over millennia then a genetic resistance might develop. This appears to have been the case almost uniquely in West Africa where the frequency of the S hemoglobin allele, which leads to sickle cell trait, increased genetic resistance to malaria (Livingstone, 1958). More often, however, stresses are not so persistent and a genetic "solution" is never developed. In these cases some degree of resistance and adaptation may derive from shorter term developmental or physiological changes (Thomas *et al.*, 1979).

Unfortunately, chronic stressors and limiting resources often overburden developmental and physiological systems. Thus, while such systems are initially able to meet environmental challenges, the persistence of the conditions often results in a decreased health and nutritional status, which will be further lowered by exposure to new stressors. Individuals who have poor nutrition, for example, are less resistant to infectious diseases, and infectious disease further lowers nutritional status (Martorell, 1980; Allen, 1984; Mata *et al.*, 1971).

The severity and duration of the stress response may be viewed as a function of the degree of cultural and environmental constraints and stressors, balanced against the adequacy of the cultural buffering system and individual resistance resources. Fortunately for the paleopathologist, the stress response, sterotypic physiological changes resulting from the struggle to adjust, is frequently manifest in relatively permanent skeletal changes. While some stressors, if rapid and system specific, may not leave any evidence of a skeletal response (Ortner, 1990), many stressor do leave scars on skeletal elements. Because bone and teeth are limited in their response repertoire, one can read that a struggle to adapt has occurred, although less in known of the stressor that was the cause of the struggle.

Health and adaptation have significance that extends beyond the individuals to the population and society. As we have outlined in the model, stresses effect populations and groups by decreasing work capacity, decreasing fertility, increasing morbidity and mortality, and causing secondary disruptions to the social, political, and economic structure of a community (Allen, 1984). In turn, these cultural and population level changes cycle back to cause further changes in the environment and cultural system.

The evolution of this model from the early to late 1980s reflects an expansion in thinking about the role of health and its significance in human adaptation. In the earlier model, the end result of a failed struggle to adapt was the death of the individual (Huss-Ashmore *et al.*, 1982). In the revised model a further step was added to demonstrate how the individually based indicators of stress, the paleopathological data, might be used to make inferences at the population level (Goodman and Armelagos, 1989). As well, the feedback arrows were added to illustrate how the stressors could have a further cyclical effect on environmental and cultural systems.

It is a challenge to paleopathologists to go beyond the individual to examine how patterns of disease and disruption have significance for families, larger social groups, and cultures. Similarly, it is a challenge to go beyond concerns with proximate causes of illness (contact with infectious pathogens, insufficient diets) to examination of the broader and underlying social, economic, and political factors which determine who comes into contact with stressors and whose resources are most limited. It is proposed that it is only in this way that paleopathology will provide insights into adaptation in the past. It is only a superficial paradox, however, that the inferences upon which these insights need to be based need to come at least in part from studies of contemporary groups.

Nutrition, Disease, and Function: The Significance of Small Body Size

Simple and uncomplicated anthropometric measurements are a powerful means for assessing a population's adaptation to environmental conditions. Except for a very limited number of ethnic groups such as Nilotic Africans, there is surprisingly little variation in mean height among different ethnic groups that can be attributed to genetic differences (Dietz, 1983; Graitcer and Gentry, 1981; Habicht *et al.*, 1974). Thus, it is now accepted procedure for nutritional scientist to compare heights in different groups against a universal standard (Habicht *et al.*, 1974). Because of the universality of the growth pattern at the group level, it is possible to compare the prevalence of stunting in groups, defined in various ways such as the percent of individuals below the 5th or 10th percentile, below negative two standard deviations (near the 2nd percentile), or below 80% or 90% of the mean of the standard. One should, for example, expect about 5% of individuals to be below the fifth percentile height-for-age due to random processes. If the observed percent is significantly greater than 5% then one can usually infer an environmental cause. In this way, the pattern of stress among and within groups can be readily compared. In the analysis of the prevalence of stunting, nutritionists and international health researchers have a powerful tool for targeting populations at risk.

While the general procedure for evaluating stunting has never been seriously challenged, in 1980 the economist David Sekler challenged a number of fundamental assumptions about the *meaning and consequences* of short stature in developing countries. Sekler proposed that most individuals that are short in stature in developing countries are short because they have *adapted* to marginal food availability (Sekler, 1980, 1982). Sekler asserted that the body adjusts to chronically low nutrient availability by reducing its growth. In this way it is able to regain physiological equilibrium. Sekler further suggests that this readjustment occurs without adaptive cost. These individuals are essentially "small but healthy" (Sekler, 1980:225). In fact, small individuals may be better adapted than larger individuals to an environment in which food scarcity is a chronic problem.

Sekler's "small but healthy" hypothesis has attracted a great deal of attention and engendered much debate among policy makers, nutritionists, anthropologists and others concerned with the international dimension of malnutrition. The main question concerns whether or not these individuals are, indeed, small but healthy. If Sekler is correct in his proposition then it follows that there are few functional inferences that can be made based on anthropometric data and we need not worry so much about the mild-to-moderately malnourished.

Contrary to the "small but healthy" hypothesis, Martorell, in an early review of the literature on growth, nutritional status and infectious disease, found that poor growth status is almost invariably associated with increased disease in developing countries (Martorell, 1980; also see Tompkins, 1988). Although this relationship appears to be somewhat attenuated in developed countries, the picture of nearly all past societies is more akin to the situation in contemporary developing than developed countries. The relationship between undernutrition and disease in developing countries is so strong that it is difficult in many cases to ascertain the cause of poor growth status (Martorell, 1989). Growth status seems to be a sign of the cumulative stress experienced by an individual, of which poor diet and disease are likely to be of main importance. A review of the literature clearly shows that the primary proximate causes of short stature are undernutrition and disease. To then affirm that short stature is adaptive implies a sanctioning of the presence of the conditions leading to short stature (Martorell, 1989), which might ultimately be traced back to poverty and inequality in access to resources (McKeown, 1988).

In addition to increased morbidity, poor growth status is associated with a number of other functional consequences. Chavez and Martinez (1982) have systematically addressed the functional cost of undernutrition and small body size in children from Tezonteopan, Mexico. Their long term nutritional supplementation study began in the 1960s. Pregnant mothers and their infants were divided into two groups of which one received daily nutritional supplementation and the other group did not. These authors found that the better nourished children cry less and talk at an earlier age. The amount of activity is highly reduced in the shorter children. The shorter and non-supplemented children displayed less exploratory behaviour (Chavez and Martinez, 1982). This study shows that growth competes with activity a limited nutrient supply. When growth is compromised activity is not spared. Rather, the reverse seems to be true. Poor growth suggests that activity has also been affected. There are functional cost in terms of decreased activity and learning that are signalled by growth stunting (Figure 4.1).

It is also clear from a number of studies that poor growth is related to an increased risk of mortality, the ultimate indicator of adaptive failure (see Alam *et al*, 1989; Chen *et al*., 1980; Lerberghe, 1988; Smedman *et al*, 1987). Alam and co-workers (1989) for example, have recently found that children in the area of Teknaf, Bangladesh with small arm circumferences suffer a twelve-fold increased probability of dying compared to children with more adequate arm circumferences. Lerberghe (1988) sums up his recent review of the relationship between childhood stunting and mortality in stating: "If both stunting and mortality are different outcomes ... of a succession of stresses on children's health, then the measure of the height of children in a community is an even more valuable tool for evaluation." (1988:259). He then goes on to say: "...the measurement of the prevalence of stunted children then becomes an operationally fairly feasible, be it indirect, measurement of poverty related disease frequency ... The prevalence of stunted children in a community appears to be a good overall indicator of the health status of a community of children..." (1988:259).

Martorell cogently summarizes the problem with the small but healthy hypothesis by asserting the following four points. First, individuals, children and adults, in developing countries are short due to chronically poor diets and frequent infection during development. To consider stunting to be adaptive is to insinuate that its causes are desirable. Second, growth retardation, "rather than an innocuous response to environmental stimuli, is a warning signal of increased risk of morbidity and mortality." (Martorell, 1989:15). Third, the conditions which cause stunting also affect other functions such as cognition and work capacity. Finally, stunted girls who grow up short are at increased risk of delivering growth retarded infants with a greater chance of dying (Martorell, 1989:15).

In considering how Sekler could have made such a nearsighted interpretation of the adaptive consequences of stunting, Pelto and Pelto (1989) suggest that Sekler fell victim to what they identify as the "quitting early" problem (also see Harris, 1988). They propose that many social scientists are guilty of frequently quitting early, before they have evaluated all effects, and especially the long-term causes and consequences of "adaptations." Sekler is clearly guilty of quitting early. Further, the claim of a "no-cost adaptation" (Beaton, 1989) suggests too much of an approval of the status quo.

What are the implications of the study of stunting and the disproving of the "small but healthy" hypothesis for studies of adaptation and health in past societies? The implications that can be drawn from data on the prevalence of stunting in contemporary populations suggests that similar data might also be a powerful indicator of group adjustment in prehistoric societies. We and others have noted some of the difficulties in studying growth patterns from skeletal remains (Goodman *et al.*, 1984; Larsen, 1987). Some of these problems include fragmentary and small sample sizes and an inability to precisely age individuals. On the other hand, long bone lengths can be accurately measured and compared to a modern standard (Maresh, 1955). Inferences can be made from the simple prevalence of subadults and adult who fall, for example, two standard deviations or greater below the mean. On balance, the number of clear and powerful inferences that can be gained from an analysis of growth status far outweighs the technical difficulties (Martin *et al.*, 1990). Examined in the context of the model (Figure 4.2) and the data on the cause and consequences of stunting in contemporary populations, growth data can be used to make strong inferences about group level adaptation.

The more general implication that can be derived from the "small but healthy" debate is that many paleopathologists also quit too soon. We quit before systemically analyzing the implications of individual disease patterns on the group and society. In a contemporary context, Chavez and Martinez (1982) have shown that moderately malnourished children may be sick as much as half the time during their childhood. It is difficult to comprehend the burden this level of illness must place on a mother in a developing community (Leatherman, 1987), and there is no reason to suggest that this burden of illness was any less in most past societies. Too often we ignore questions about the burden of illness on the functional and adaptive capacities of individual and groups.

Ortner (1990) has voiced one interesting challenge to not quitting early. His concern is that we think carefully about the meaning of infectious lesions on bone. Ortner suggests that these lesions should not be interpreted as a sign of stress, but rather as signs of rallying from insults, indicating an adaptive response to infection. Just as the raising of the question was important for nutritionists who rallied to consider the validity of the "small but healthy" hypothesis, the raising of the question concerning the adaptive significance of skeletal indications of disease is an important call to paleopathologist to think through their data. I suggest further that Ortner's challenge has to go even further to not just thinking through the individual response to the insult, but the systemic effect that the struggle to adapt to an infectious disease might have on a family and larger social groups.

A number of research initiatives can strengthen to a greater extent inferences on the functional and adaptive consequences of our indications of health. In some cases it might be important to know the consequences of an indication of stress in a contemporary context. An example of this concerns enamel hypoplastic defects. In order to make clearer inferences about the meaning of enamel hypoplasias, we need to better know the sensitivity and specificity of this indicator to bouts of malnutrition and infection (Goodman and Rose, 1990).

In other cases the underlying cause as well as the functional consequences of a stressor are known. For example, it is clear that most cases of porotic hyperostosis are secondary to iron deficiency, and the effects of iron deficiency on function are increasingly known from contemporary studies (Chwang *et al.*, 1988; Dallman, 1987; Pollitt, 1987). In this case, research is needed on the precise relationship between the skeletal indicators of iron deficiency and biochemical indicators. How chronic and severe a deficiency is indicated by iron deficiency? Although similar knowledge "gaps" are found for most paleopathological measures, I remain optimistic about the ability of well directed research efforts to fill these gaps.

Conclusions and Recommendations for Future Research

There are three fundamental points that I hope to have underscored in this discussion of our ability to understand health and its relationship to adaptation in past societies. The most general and theoretical point concerns the synergetic and interactive relationship between health and adaptation. Health is a function of the organism's ability to adjust to environmental constraints and stressors. It is the result of the adaptive process. Poor health constrains adaptation. It is a condition that needs to be alleviated or adjusted to. In this sense, health, classically defined in terms of morbidity and mortality, is a measure of the success of adaptation.

Under favourable conditions, adaptation is manifest in good health, and health furthers one's ability to adjust to stressful conditions. Such conditions can increase one's adaptive flexibility and ability to endure other stressors and constraints. Conversely, the inability to adapt is apparent in poor health, which in turn decreases one's ability to adapt to further insults. Stressful environmental conditions promote health problems, which constrain adaptation by, among other things, lowering resistance to disease, lowering work and productive capacity, lowering interest in and ability to engage in social and discretionary activities, interfering with learning and cognitive functioning, decreasing fertility and fecundity, and interfering with nutrient intake and utilization (Allen, 1984; Buzina *et al.*, 1989). The process by which health is related to environment and adaptation is nothing less than intrinsic, interactive, and cyclical.

The second fundamental point is that health is a measurable entity in studies of contemporary as well as past societies. Readily available measures of health status, such as rates of infant mortality and stunting, are routinely used to gauge and compare the general level of health of contemporary groups (Goodman *et al.*, 1988). While health status, as well as stress and nutritional status, is a behavioral phenomena that can not be directly measured in past societies, there are numerous commonly employed and increasingly refined methods for reconstructing health, stress, and nutritional statuses from bones and teeth (see for examples the following reviews: Buikstra and Cook, 1980; Goodman *et al.*, 1984; Gilbert and Mielke (eds.), 1985; Larsen, 1987). Some of these measurements, such as those involving mortality and growth rates, are particularly robust in their ability to generate adaptive inferences because there are equivalent and commonly used contemporary measures (Goodman *et al.*, 1988).

While taphonomic processes render it impossible to reconstruct fully all dimensions of health experienced by an individual or a group -- either in the past or currently -- I submit, optimistically but not without empirical and scientific justification, that many techniques are now available that can provide an accurate appraisal of a number of dimensions of health in past populations. The information derived from these techniques, coupled with an understanding of the adaptive and functional significance of the measures of health, can provide key insights into the dynamics of adaptation in past societies.

Thirdly, studies of health in past societies can provide unique insights into the dynamics of past societies and the evolution of the human condition. Unfortunately, many paleopathologists have quit too soon, before the significance of health has been fully explored.

In concluding I wish to make reference to the theoretical underpinning of the model and perspective outlined above in relationship to recent theoretical challenges in archaeology. It is somewhat paradoxical that the research paradigm which is highly processualist in nature is proposed at a time when archaeology seems to be moving away from a processualist paradigm (for example see Hodder, 1986). This paradox may be explainable by the fact that one of the processualist paradigm's greatest failings, lack of concern for individual and culturally specific dynamics, is far less applicable to biological phenomena. The biological impact and consequences of health are more nearly universal than social processes. Therefore, contemporary analogies about the cause and effect of health are quite consistent and can be used to make assumptions about cause and effect in the past. Further, an examination of health can and should be seen not as isolated local-level phenomena, but as a result of the openness of local systems with other systems. In fact, the modern ethnographic data suggest that poor health is invariably tracked back to poverty, which is itself due to regional and interregional processes. The fault of the new archaeology of over emphasis on local ecological processes need not be shared by a processualist view of health and adaptation. Thus, it is not paradoxical that when some social archaeologists are about to discard processes, middle range theory, and ethnographic analogy, bioanthropologists may discover the usefulness of these methodological and theoretical underpinnings. The study of health in past societies needs to be placed in an evolutionary and ecological context, linking it to studies of contemporary groups. In this way we will better understand ideas about progress, the dynamics of past societies, and the evolution of the human condition.

References

Alam, N., Wojtyniak, B. & Rahaman, M. (1989). Anthropometric indicators and risk of death. *Am. J. Clin. Nutr. 49*:884-88.

Allen, L. (1984). Functional indicators of nutritional status of the whole individual or the community. *Clin. Nutr. 35*:169-75.

Beaton, G. (1989). Small but Healthy? Are we asking the right question? *Hum. Org. 48*:30-39.

Buikstra, J. and Cook, D. (1980). Paleopathology: an American account. *Ann. Rev. Anthropol. 9*:433-70.

Buzina, R., Bates, C., van der Beek, J., Brubacher, G., Chandra, R., Hallberg, L., Heseker, J., Mertz, W., Pietrzik, L., Pollitt, E., Pradilla, A., Suboticanec, K., Sandstead, H., Schalach, W., Spurr, G. & Westenhofer, J. (1989). Workshop in functional significance of mild-to-moderate malnutrition. *Am. J. Clin. Nutr. 50*:172-76.

Chavez, A. and Martinez, C. (1982). *Growing Up in a Developing Country*. Mexico: Instituto Nacional de la Nutricion.

Chen, L., Chowdhurry, A., & Hoffman, S. (1980). Anthropometric assessment of energy-protein malnutrition and subsequent risk of mortality among preschool aged children. *Am. J. Clin. Nutr., 33*:1836-45.

Chwang, L., Soemartri, A. & Pollitt, E. (1988). Iron supplementation and physical growth of rural Indonesian children. *Am. J. Clin. Nutr. 47*:496-501.

Dallman, P. (1987). Iron deficiency and the immune response. *Am. J. Clin. Nutr. 46*:329-34.

Dietz, W. (1983). One for all? thoughts on reference standards for growth in short populations. *Nutr. Res. 3*:129-31.

Gilbert, R. and Mielke, J. (1985). eds. *The Analysis of Prehistoric Diets*. Orlando: Academic Press.

Goodman, A. and Armelagos, G. (1989). Infant and childhood morbidity and mortality risks in archaeological populations. *World Archaeol. 212*:225-43.

Goodman, A., Martin, D., Armelagos, G. & Clark, G. (1984). Indications of stress from bones and teeth. In *Paleopathology at the Origins of Agriculture*. ed. M.N. Cohen and G.J. Armelagos. pp. 13-49. Orlando: Academic Press.

Goodman, A. and Rose, J. (1990). Dental enamel hypoplasias as indicators of nutritional status. In *Advances in Dental Anthropology*. eds. M. Kelly and C. Larsen. AR Liss: New York (In press).

Goodman, A., Thomas, R., Swedlund, A. & Armelagos, G. (1988). Biocultural perspectives on stress in prehistoric, historical, and contemporary population research. *Yearb. Phys. Anthropol. 31*:169-202.

Graitcer, P. and Gentry, E. (1981). Measuring children: one reference for all. *Lancet 2*:297-99.

Habicht, J.-P., Martorell, R., Yarbrough, C., Malina, R. & Klein, R. (1974). Height and weight standards for preschool children. How relevant are ethnic differences in growth potential? *Lancet 1*:611-15.

Harris, M. (1988). Cultural materalism: alarums and excursions. In *Waymarks: The Notre Dame Inaugural Lectures in Anthropology*. ed. K. Moore. pp. 107-26. Notre Dame: University of Notre Dame Press.

Hodder, I. (1986). *Reading the Past*. New York: Cambridge University Press.

Huss-Ashmore, R., Goodman, A. & Armelagos, G. (1982). Nutritional inference from paleopathology. In *Advances in Archaeological Method and Theory Vol. 5*. ed. M. B. Schiffer. pp. 436-74. Orlando: Academic Press.

Larsen, C. (1987). Bioarchaeological interpretations of subsistence economy and behaviour from human skeletal remains. In *Advances in Archaeological Method and Theory Vol. 10*. ed M. B. Schiffer. pp. 339-445. Orlando: Academic Press.

Leatherman, T. (1987). *Illness, Work and Social Relations in the southern Peruvian Highlands*. Unpublished Doctoral Dissertation, University of Massachusetts, Amherst, Department of Anthropology.

Lerberghe, W. (1988). Linear growth retardation and mortality. In *Linear Growth Retardation in Less Developed Countries*. ed. J.C. Waterlow. pp. 245-64. New York: Raven Press.

Livingstone, F. (1958). Anthropological implications of sickle cell gene distribution in West Africa. *Am. Anthropol. 60*:533-62.

Maresh, M. (1955). Linear growth of long bones of extremities from infancy through adolescence. *Am. J. Dis. Child. 89*:725-42.

Martin, D., Goodman, A., Armelagos, G. & Magennis, A. (1990). *Black Mesa Anasazi Health: Reconstructing Life from Patterns of Disease and Death*. Carbondale: Southern Illinois University Press (In press).

Martorell, R. (1980). Interrelationship between diet, infectious disease, and nutritional status. In *Social and Biological Predictors of Nutritional Status, Physical Growth and Neurological Development*. eds. L. Greene and F. Johnston. pp. 81-106. Orlando: Academic Press.

Martorell, R. (1989). Body size, adaptation and function. *Hum. Organ. 48*:15-20.

Mata, I., Urrutia, J. & Lechtig, A. (1971). Infection and nutrition of children of a low socioeconomic rural community. *Am. J. Clin. Nutr. 24*:249-59.

May, J. (1960). The ecology of human disease. *Ann. N.Y. Acad. Sci. 84*:789-94.

McKeown, T. (1988). *The Origins of Human Disease*. New York: Blackwell.

Ortner, D. (1990). Theoretical and methodological issues in paleopathology. In *Human Paleopathology: Current Syntheses and Future Options*. eds. D. Ortner and A. Aufterheide. Washington, D.C.: Smithsonian Institution Press (In press).

Pelto, G. and Pelto, P. (1989). Small but Healthy? An Anthropological Perspective. *Hum. Org. 48*:11-15.

Pollitt, E. (1987). Effects of iron deficiency on mental development: methodological considerations and substantive findings. In *Nutritional Anthropology*. ed. F. Johnston. pp. 225-54. New York: Alan R. Liss, Inc.

Sekler, D. (1980). Malnutrition: an intellectual odyssey. *West. J. Agric. Econ. 5*:219-27.

Sekler, D. (1982). Small but healthy? Some basic problems in the concept of malnutrition. In *Newer Concepts in Nutrition and Their Implications for*

Policy. ed. P.V. Sukhatme. pp. 139-48. Pune, India: Maharashtra Association for the Cultivation of Science.

Smedman, L., Sterky, G., Mellander, L. & Wall, S. (1987). Anthropometry and subsequent mortality in groups of children aged 6-59 months in Guinea-Bissau. *Am. J. Clin. Nutr. 46*:369-73.

Thomas, R.B., Winterhalder, B. & McRae, S.D. (1979). An anthropological approach to human ecology and adaptive dynamics. *Yearb. Phys. Anthropol. 22*:1-46.

Tompkins, A. (1988). The risk of morbidity in the stunted child. In *Linear Growth Retardation in Less Developed Countries*. ed. J.C. Waterlow. pp. 185-99. New York: Raven Press.

Figure 4.1 A peasant family from Tezonteopan, Mexico. The two older girls are enrolled in the study of Chavez and Martinez (1982). The one with the white t-shirt with the rose (Rosa) is nineteen and was not nutritionally supplemented. The other sister is fifteen years old and has been given nutritional supplements since birth. The five inch height difference between these same sex sibling is typical of the differences between the supplemented and non-supplemented children. In addition to the height differences, Chavez and Martinez (1982) have found profound differences in activity levels, sociability, and resistance to disease.

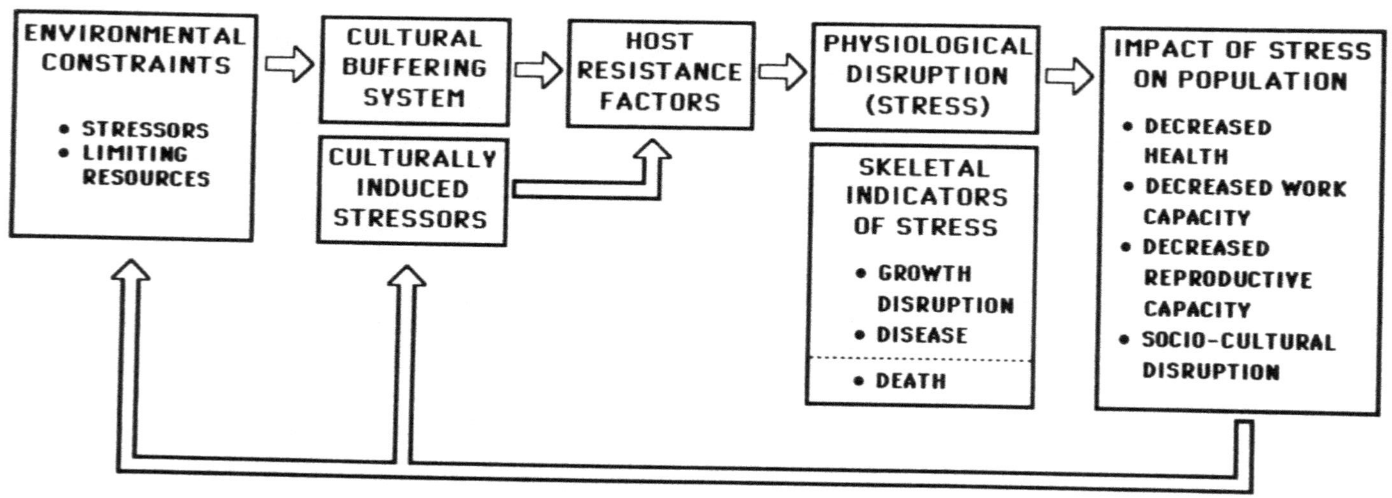

Figure 4.2 A general model for studying the relationship between health, stress, and adaptation in skeletal populations (from Goodman and Armelagos, 1989).

5. DIRECT GENETIC AND IMMUNOLOGICAL INFORMATION IN THE RECONSTRUCTION OF HEALTH AND BIOCULTURAL CONDITIONS OF PAST POPULATIONS: A NEW PROSPECT FOR ARCHAEOLOGY.

Cristina Cattaneo

Health and Archaeological Reconstruction

In recent years, the increasing interest in the study of organic remains (botanical, faunal, human) retrieved from archaeological contexts has brought forth a serious attempt to integrate information obtained from biological investigations with that of a more typically archaeological nature such as, for example, material culture. However, regardless of these new approaches, the biomedical field has not yet been able to find its deserved niche within the discipline of archaeology.

Previous studies in palaeopathology have, in fact, confronted the challenges of unravelling the history and evolution of specific diseases, of diagnosing pathological conditions in individual skeletons, and of using the most modern biomedical laboratory techniques to extract biomolecules such as haemoglobin, albumin, collagen and DNA from human skeletal material. All these may be very interesting achievements *per se* but, when isolated from a biocultural context, they are useless in the reconstruction of lifestyle and dynamics of past populations.

The state of health in particular should be considered crucial information by the archaeologist, especially when assessed at a population level, for it is intertwined tightly with phenomena such as population dynamics, lifestyle and culture change. From the health conditions of a population or of subgroups among that population, one can begin to make inferences concerning subsistence, nutrition, class differences, and so forth. For example, when among a skeletal population certain subgroups seem to have been affected (to a greater degree than other groups) by "skeletally diagnosable" stress factors, or seem to have suffered from nutritional deficiencies, though material culture may not give us any clues concerning this matter, we can, on the basis of skeletal evidence only, begin to formulate the hypothesis that these individuals belonged to a less privileged social class. In other words, a closely knit interaction between the study of culture and that of human biology is essential.

To reconstruct the health status of a population it is first necessary to understand the interaction between man as a biological entity and the environment surrounding him. It is also important to acquire a proper understanding of the causative factors behind good or bad health conditions. Further, it is crucial to realise that there are two equally relevant main contributing factors to disease: one is the environment, the other, not less important, is genetic predisposition to disease (Figure 5.1).

Health: Genetics or Environment?

Under the heading "environment" it is reasonable to include an array of phenomena which may influence the individual from the "outside". This array may range from micro-organisms, a frequent cause of disease, to climatic conditions, economy and even cultural habits, which may affect the lifestyle of an entire population (a certain degree of circularity is, of course, inevitable when the reciprocal effects of health/biology and culture are considered). On the other hand the term "genetic predisposition" refers to the innate predisposition of individuals (or populations) to develop a particular disease. Knowing the nature of these predispositions is extremely advantageous, since all diseases arise as a result either of genetic susceptibility, environmental factors or, many times, of these two phenomena combined.

It is precisely the genetic component of disease that I wish to elucidate in this article. The aim of this paper is to explain the relevance, not only in the medical but also in the archaeological context, of the genetic contribution to some diseases; to show how this information may help us to reconstruct more accurately the health status and lifestyle of a population, and to suggest how this genetic data may be obtained from specific biological molecules retrieved directly from skeletal material (see below).

So far, we have seen that to make an accurate statement concerning the lifestyle and dynamics of a population one must understand its health status, and to do this, one must realise how extensively environment and genetics contribute to disease. Therefore, to reconstruct health status, not only should we look for the environmental factors which may affect it, but we must also try to determine what and how much was contributed by genes alone. This kind of information may help avoid incomplete or erroneous reconstructions of health based only on skeletal lesions and demographic patterns. In other words, one should try to comprehend whether individuals were destined, because of some genetic characteristic or anomaly, to develop specific conditions (or to be more susceptible to acquiring diseases), or if environmental factors were the agents chiefly responsible for the onset of a particular disease. We should avoid pitfalls such as attributing the causes of illness or bad health to environmental factors when the genetic makeup is primarily responsible. An example of this kind of error is seen when considering cases of anaemia diagnosed according to skeletal lesions. At times, the palaepathologist is confronted with skeletal evidence of anaemia in the form of porotic hyperostosis (thickening and porosity of the cranium and of the roof of the orbits). On occasion, these lesions could have had a genetic basis (as in the case of thalassemia or Mediterranean anaemia). Other times, they could have been a consequence of adverse environmental factors, such as iron deficiency. Many palaeopathologists (e.g. Steinbock, 1976; Ortner and Putschar, 1985) try to differentiate between the skeletal manifestations of iron deficiency anaemia and those of genetic anaemias. But this is a difficult task, simply because the changes in the bone tissue are too non-specific (Hart, 1980). In practice, these lesions often have been hastily attributed to dietary iron deficiency, and the hypothesis that the condition may have been genetic is set aside. This type of error could be excused if the implications of such a diagnosis were not so important to the archaeologist. In fact, attributing the anaemia to dietary deficiency will lead to the idea that the population, or parts of it, lived on an inadequate diet, whereas the disease, being genetic, may have been inevitable, even under the best of nutritional conditions. Angel (1967) also has recommended being very careful of this pitfall. In skeletons from the early Neolithic (c. 6500 B.C.) and Bronze Age Greece, one can notice these lesions in the bones of both children and adults. Angel has related this to the presence of endemic malaria and therefore, indirectly, to the presence of thalassemia. Thalassemics in fact seem to be protected, to a certain degree, against malaria. As evolutionary theory would have it, individuals carrying a "degree" of thalassemia compatible with life would have been better adapted and therefore selected for in such an environment. In this context it is not possible, according to Angel, to make any dietary implication. On the other hand, if one were studying a population known not to carry the gene anomaly for thalassemia, such environmental inferences could be made. Thus, knowing the genetic predisposition to disease of individuals may help in the reconstruction of the past. The need to focus on the possible existence of an innate tendency to develop or acquire a specific disease must not be under-estimated, nor should the possibility of retrieving this information directly from skeletal material. How, then, should we obtain this kind of information?

Sources of Genetic Information on Disease

So far, two solutions to this problem have been suggested. One is to look for signs (morphological traits) on the skeleton which seem to be associated with susceptibility to disease (see below); the other is to try to determine the racial affiliation of the population studied. Sometimes, in fact, the identification of the racial group to which the individuals studied belong may indicate particular predispositions to disease. This is done on the basis that some racial groups are known to be more likely to develop certain diseases than others. Finally, a third solution (the one suggested in this paper) considers the possibility of obtaining this information from biomolecules which may be present still in the bone material itself.

At this point it would seem convenient, since genetics and race are terms that will be encountered frequently, to define what is meant by race in the current context. For our purposes, a race can be defined as a distinct group of individuals which distinguishes itself from other populations by a gene pool in which gene frequencies are characteristic of that population and of no other (Thompson and Thompson, 1966). In other words, races will distinguish themselves as differing in some of the genes they carry. This is a purely "genotypic" definition, as opposed to a "phenotypic" one. Whereas the genotype is the genetic constitution itself, composed of different genes which are in turn "made of" DNA, the phenotype is the ensemble of easily seen external characteristics. The phenotype may reflect closely the genotype, but it may also be the result of "environmental" factors which influence and somewhat alter the expression of the genotype. So the phenotype does not necessarily exhibit the genotype (Davies, 1986). The genotype, on the other hand, without external biases, consists of the genetic connotations of each single person. So even if in some authors' definitions phenotypic characteristics come into the description of race, it is safer, when trying to differentiate groups and when studying population origins, movements and mixing, to rely on genetic information which may not be immediately noticeable morphologically but which comes directly from the genes themselves. Such information can only be obtained by studying the genetic material itself (DNA) or molecules closely related to it, as we will see later. This definition and this entire issue are particularly relevant to archaeology and anthropology. In fact, when dealing with human skeletal material, in the past genetic affiliations among individuals and the definition of race in general were greatly based on morphological characteristics of the skeleton. The inferences on racial affiliation in this case are not very reliable, due to the fact that, except for very obvious and major morphological differences, it is not yet known how, if at all, and to what extent, genes are responsible for these characteristics. If it were possible, therefore, to obtain "genetic" information directly from DNA, or from closely related molecules, then the definition of genetically related groups would be more reliable.

After this necessary digression it is now possible to return to the methods of studying the predisposition of individuals to disease. Concerning the first method, studies of cranial morphology have been performed in an attempt to observe any correlation with specific diseases. A study by Olivier *et al* (1975) on a group of nineteenth century skeletons from Lisbon points out that there seems to be an interesting relationship between skull form and differential mortality by tuberculosis, plague and smallpox. Subjects who had died from these diseases had a longer head and face. The practical applications of this system are very limited, however. As far as the second method,racial affiliation, is concerned, various studies have been performed on the morphological and metric characteristics of the crania of different races and on how such data may help differentiate between groups of populations. This, in turn, could in some instances indicate which individuals are more likely to contract specific diseases. But how reliable is any kind of information obtained from gross morphological characteristics of the cranium or of any other part of the skeleton?

The use of osteological traits as a source of genetic/disease-related information (with the exception of pathological lesions, of course) can be misleading and inconclusive, the main argument against this procedure being that it is not yet known how metric or morphological traits are inherited nor how they develop. Schwidetzky (1979) is correct in his observation that many of these characteristics are not age-stable, have a continuous variability and that the conditions for the development of these traits has yet to be discovered. Therefore, though they may have some poorly understood significance, these traits cannot provide a reliable source of genetic information, despite Berry's (1979) attempts to make them seem relevant. Some researchers, Buikstra (1976) for example, have looked at osteological traits at a population level with the precise intention of integrating them with other archaeological data, thus presenting an excellent example of inter-disciplinary blending which should be emulated. Even so, osteological traits as a source of genetic information of any sort may not be as reliable as expected. Other sources of this kind of information must be pursued.

Where, then, should one turn for reliable genetic information obtained directly from skeletal material? The answer lies in biologically informative molecules that may survive in ancient bone, in particular DNA and certain types of proteins which are a direct "product" of genes. If it were possible to retrieve relevant biomolecules from human osseous archaeological material they would certainly provide, in theory, a much more reliable source of genetic information. Before discussing the particular types of biomolecules which can give us this kind of information it would be sensible to consider how feasible their extraction is from ancient skeletal material.

The Extraction of Informative Biomolecules

Very little is known concerning the diagenesis and fate in general of organic material within bone after hundreds or thousands of years. Studies have described how cells die and how organic material degrades, but little has been said about the influence of time, soil conditions, temperature, amount of oxygen, leaching effects of water and so forth on biological molecules. Proteins and other biomolecules have been isolated from ancient skeletal material. Unfortunately, as Hedges (1978) has pointed out, the components of bone that seem to survive in great abundance are the ones which are less informative and less suitable for acquiring information on disease susceptibility and genetics. In fact, most relevant molecules are quite soluble and perhaps more easily lost to the effects of leaching, and are also more accessible to attacks from micro-organisms in the ground which may alter their nature. This last factor seems to be one of the main reasons why studies on the possibility of retrieving molecules of the ABO blood groups (ABO blood group antigens) are so ambiguous and unreliable. A second problem is that the organic residues which do persist in time will certainly be present in very small quantities. This obstacle may be overcome by the use of modern immunological laboratory techniques capable of detecting molecules present in nanogram (10^{-9}) to picogram (10^{-12}) amounts.

Given these difficulties, it might seem very unlikely that organic material will survive in any type of ancient bone (or artefact) but suprisingly, many biomolecules such as collagen, a substance found in most tissues, have been isolated by Gurtler *et al* (1981). Loy (1983) has discovered species-specific haemoglobin crystals from 2000 year old blood-stained arrow heads and has also identified human blood from prehistoric rock paintings (Loy *et al*, 1990). Using radio-immunological assays, Lowenstein (1980, 1981, 1985) has detected albumin (a protein found in the blood) in 1.9 million year old fossils. He has also found species-specific proteins in Neanderthals, *Homo erectus* and *Australopithecus robustus* (1981). Ascenzi *et al* (1985) have detected immunologically alpha and beta haemoglobin chains in bone up to 4500 years old, and Rothschild and Turnbull (1987) seem to have detected, via immunohistological techniques, antigens of the *Treponema* bacillum in lesions of a Pleistocene bear. Again with immunological methods Ulrich (1987) has extracted osteocalcin (a protein present in the mineral phase of bone) in animal fossils from 12,000 to 13 million years old. He also subjected bone to very high temperatures, as did Lowenstein when looking for albumin, and found that the antigenic activity of the molecule, its distinguishing chemical characteristics in other words, had been retained.

DNA has been detected (Jones, 1986; Paabo, 1986) in mummified tissue and recently (Hagelberg *et al*, 1989) in human skeletal remains. When looking for genetic information concerning characteristics such as race and susceptibility to disease, the availability of the DNA molecule itself, or fragments of it, would supply the most direct genetic information available and would in theory be the most reliable source of such information among most other biomolecules. But, as Paabo (1986) has pointed out, although DNA seems to be resistant to harsh agents, there appear to be problems concerning the very low yield obtained when attempting to clone ancient DNA. Venanzi and Rollo (1990) have even suggested that the nucleic acids (components of DNA) in ancient mummies are mainly of RNA (another type of genetic material, different from DNA). Although Hagelberg *et al* (1989) have succeeded in isolating DNA from skeletal material, further research must be done in order to confirm this, and especially to control for the possibility of contamination from fresh DNA, seeing that the experiment employs an extremely sensitive technique. Nonetheless, if DNA does survive in ancient material, the implications for the fields of archaeology and anthropology would be very significant. Another group of biomolecules which would be extremely helpful in these kinds of investigations (and would be a close second to DNA in the type of genetic information they can provide) is that of the HLA antigens, which will be considered in detail further on. Although the extraction of this "glycoprotein" (i.e. HLA) currently presents difficulties even where relatively fresh material is concerned, the possibility of its survival in ancient material should be considered seriously.

Generally speaking, the causes and mechanisms of survival of these molecules are not known. A partial explanation may be provided by the particular biochemical characteristics of the molecules or, perhaps, by a certain degree of protection given by the apatite (i.e. mineral) matrix or the trabecular structure in certain parts of the bone itself. In conclusion, it has been proven that various organic molecules do survive in human osseious material.

It is important to understand how different and how delicate this kind of research is compared to that concerning skeletal morphology and skeletal lesions. The search for any kind of biomolecule implies looking for something one cannot actually see, and therefore it consists, first of all, of extracting from the bone the substance one hopes is present and then using an adequate test to detect its presence. The tests may vary according to the molecule which is being sought, and may be more or less complex. But the logic behind this kind of research is, generally speaking, the same regardless of what one is seeking. They are tests, or assays, which have been developed in the field of biomedical sciences. As an example one can consider trying to extract a fairly simple molecule: the blood protein albumin. This procedure consists, first of all, of taking perhaps 10 g. of bone from the vertebrae, femoral head or, perhaps, the iliac crest. This bone should be crushed and then, for a period of time, put into a solution which should help extract the albumin from the bone. After this, the solution in which the bone powder was immersed is collected. Hopefully it will contain some albumin. This amount of liquid (the extract) is then concentrated to a few millilitres, for example, and tested. Finding the appropriate test with which to detect the presence of albumin in the extract is another crucial step. The technique must be both very sensitive, in order to detect very small quantities of albumin, and very specific, so that another molecule, chemically similar to albumin, will not be erroneously picked up as albumin. An example of such a test is the technique known as ELISA (enzyme-linked immunosorbent assay). This system uses molecules called antibodies, which will recognize and "stick to" only a specific molecule. Different antibodies can be made to recognize different molecules. So if a solution of antibodies which will recognize only albumin is mixed with the extract (this is usually done in very small wells of a plastic plate), and if any albumin is present in the extract, the antibody will "attach" to it. This cannot be visualised immediately, so a second antibody which will recognize the first one has to be used. This second antibody also cannot be "seen", but it has attached to it a molecule which will induce a third substance finally to cause a fourth added substance to turn a certain colour. So the process of detecting albumin consists of adding one substance after another, each one of which will attach to the previous one until, finally, the addition of a colouring agent will allow us to say whether albumin is present in that particular bone extract or not. It is obvious that the logic to follow in performing these experiments, and the series of difficulties one has to confront, are novel to the field of archaeology.

The remaining paragraphs of this chapter concern the above-mentioned HLA antigens. The choice of focusing on these proteins was dictated by their capacity, perhaps more than that of any other biomolecule except for DNA, to supply genetic information. The decision to focus on HLA instead of DNA was made partly because of the personal experience the author has in studying and attempting to extract HLA from ancient material, partly because it could prove to be an alternative method (with some advantages and, equally, some disadvantages over DNA, but which has not received as much attention) for obtaining genetic information.

HLA Antigens as a Source of Genetic Information

If we are to look for genetic information, we must choose a molecule which will reflect best the information contained in our genes. One candidate is the complex biomolecule called HLA antigen (Human Leukocyte Antigen), which is anchored to the surface of most cells in our body and which regulates our immune response. The entire HLA system, whose counterpart in our genes is called the Major Histocompatibility Complex, is the most remarkable genetic marker. It lends itself favourably to studies, not only of selection, human evolution and genetics (Bender, 1984) (for paternity tests and genetic characterisation of different human populations, for example) but it also, and especially for our purposes, lends itself to disease association studies.

a) Biology of HLA antigens

The more complex details of this molecule are beyond the scope of the chapter. Nonetheless, a brief description of its structure and function may perhaps help to understand its importance. The Major Histocompatibility Complex (MHC) is a complex of genes which codes for the production, by the cell, of HLA antigens. In man, HLA antigens were called Human Leukocyte Antigens because they were first seen on lymphocytes, a type of white blood cell. These molecules are proteins, more precisely glycoproteins, which are expressed and embedded within the surface of the cells of all higher vertebrates and whose function is to regulate the immune response which is crucial to the survival of individuals (Bender, 1984) (Figure 5.2).

Briefly, genes (DNA) provide information which the cell may translate into proteins. The information sent by the genes orders the cell to string together a series of "building block" molecules called aminoacids to produce a protein molecule such as an HLA antigen. Different combinations of a limited variety of aminoacids and different lengths of these strings will allow for different kinds of proteins (there are many more complicated functions involved in this instance, but for the purpose of this chapter, this definition will suffice).

A locus is a position on the genetic material which houses one gene. HLA antigens can be "ordered" by a number of different genetic loci. They have been separated into three classes: Class I, Class II and Class III (Schiebl, 1986). Focusing on Class I antigens, for example, one can see that they are products of loci A, B and C and that they are expressed on the surface of all nucleated cells and platelets (Bender, 1984) (because of their vast distribution, they seem to function as markers of self-recognition; in fact, only in the context of self-recognition may organisms recognize non-self and take immunological action against it). One locus can house different variants of a gene which are called alleles. These alleles will code for different "varieties" of the same protein (e.g. HLA antigen) (Figure 5.3). So locus A, for example, can code for up to 21 different types of HLA-A antigens (Thomson, 1986), of which each individual can only display two. The alleles are assigned numbers preceeded by the letter of the respective gene locus. Table 1 gives a summary of the different kinds of HLA antigens.

Each individual can express no more than two antigens for each locus. The HLA antigen combination for Class I, II, III antigens, determined by the genes of one chromosome, is the haplotype, of which every individual has two, one of each homologous chromosome. Humans, in fact, have 46 chromosomes (the structures into which genetic material is "rolled" up) which consist of 23 sets of two homologous chromosomes, i.e. two structures containing the same kind of information, one set coming from the mother, the other from the father. The haplotype is passed on as a unit. Thus, each of the two parents will offer two haplotypes, but only one from each parent will be randomly accepted, so that from two parents, only four haplotype combinations are possible (Figure 5.4). An example of an individual's HLA type, considering only the A and B loci for the sake of simplicity, could be: A1, A11, B8, B35 or A1, A3, B7, B8 or many other combinations of A and B.

b) HLA antigens as indicators of race

The frequency of HLA types, and of specific combinations of these, is important in characterising population groups and racial affiliation. Table 5.2A shows a list of these differences, as does Table 5.2B for populations within and close to Europe. For example, HLA-A1, -A3, -B8 are found mainly among Caucasians; -A24, AW31, -B40 among orientals and AW30, -B17 and BW42 among Negroes (Bender, 1984). It is known that different areas of Europe differ in HLA frequencies, especially Northern and Western Europe from central and Eastern Europe (Bodmer and Bodmer, 1978).

c) HLA antigens as possible indicators of disease

Besides helping us to determine the racial affiliation of individuals, and therefore helping indirectly to provide information on predisposition to disease, these antigens can be, themselves, directly associated with particular diseases (Table 5.3). Association in this case means, in modern medical terms, that the patient population (Thomson, 1986) has a significantly increased frequency of a particular HLA antigen compared to a control population. Some of these associations are not very strong. On the other hand, some are quite marked, such as that between HLA-B27 and ankylosing spondylitis. Research has been carried out on these antigens to elucidate their particular relationship with infectious diseases such as smallpox, malaria, cholera, tuberculosis (Bodmer and Bodmer, 1978) and leprosy (de Vries *et al*, 1976). It is clear, therefore, that the presence of certain HLA antigens may indicate susceptibility to certain diseases. The susceptibility associated with the presence of these particular antigens may be a direct effect of the HLA antigen, but in most cases the HLA antigen is simply a marker closely linked to a locus which is causally related to the disease. The point remains that these molecules are potential markers for disease. These antigens may even give us some insight into what illnesses these populations may have faced, even if they did not leave their mark on bone, which has been the only source of evidence for disease available to the palaeopathologist up until now.

d) HLA antigens: work to date

A literature search has failed to show any thorough investigation on HLA antigens in bones apart from the attempt made by Hedges and Wallace Carmichael (1978) in which the antigens were not found. Extraction of HLA antigens from Saxon and Medieval human bone has been attempted at the Regional Blood Transfusion Centre in Sheffield (Cattaneo, 1988). The method used for testing extracts made from fresh and ancient human bone (as described previously in this article) were similar to those used in tissue typing modern individuals (i.e. determining their HLA characteristics, with a view to transplants of particular organs or tissues, for example). This test, called lymphocytotoxic test, was also employed by Stastny (1973) to test mummified soft tissue for the presence of HLA antigens. The details of the experiments performed at the RBTS are beyond the scope of this chapter. The extraction methods employed succeeded in retrieving albumin, but problems arose with HLA. The difficulty in extracting this molecule, if present, from ancient or modern bone lies especially in the double nature of the extraction. It must be retrieved both from the recesses of the cancellous bone and from the fragments of cell membrane in which it is possibly still embedded. More research needs to be done before anything may be said concerning the feasibility of extracting these molecules.

Conclusion

I hope I have succeeded in stressing the importance, as well as the difficulties and the need for further research, in acquiring genetic information on race and health directly from bone in the form of biological molecules, rather than from less reliable sources such as the study of the gross morphology of the skeleton or of blood groups or HLA antigen distribution among modern populations. In fact, biochemical and serological studies concerning genetic traits and variability have been performed on modern populations in the attempt to shed some light on human genetic population patterns of the past, by extrapolating modern data to an ancient context. However, because it is not possible to control for the great number of factors and variables which over the millennia could have affected the final genetic make-up of modern populations, "genetic" studies performed directly on ancient bone are in theory more reliable.

We should therefore direct our attention to determining what happens to these molecules and how they may be retrieved, to deciphering the information on innate susceptibility to disease (or any other genetic kind of information we may need), and then to integrating this essential background information into the rest of the archaeological picture (Figure 5.5).

A brief mention should also be made concerning the importance, in the context of disease, of looking for biomolecules that are not necessarily human; for example, molecules characteristic of specific micro-organisms that affect man and which may still be present in lesions on the skeleton of the individual. In other words, lesions caused by infectious diseases such as tuberculosis or syphilis may still contain diagnostic biomolecules of the specific bacterial agents responsible for the onset of the disease. These may allow us to make a more accurate diagnosis.

Finally, the point I would like to convey is that no matter which particular biomolecule we might consider recovering from skeletal material, be it DNA, HLA antigens or blood proteins, these genetically informative or disease-associated molecules do exist and there is a real possibility that they may persist in time. Therefore more effort should be spent in trying to understand the processes of degradation of these molecules and the conditions of their survival. Their retrieval and extraction from human skeletal material can give us some clues concerning the genetic (innate) susceptibility to disease. This would, in turn, allow a sounder base for the reconstruction of the health status of a population and, together with other archaeological data, would help draw a more accurate picture of the lifestyle and cultural changes of past populations.

References

Alberts, B., Bray, D., Lewis, G., Raff, M., Roberts,K. & Watson, J.D. (1983). *The Cell.* New York: Garland.

Angel, J.L. (1967). Porotic hyperostosis or osteoporosis symmetrica. In *Diseases in Antiquity* eds. D. Brothwell and A. Sandison. Springfield: Charles C. Thomas.

Ascenzi, A., Brunori, M., Citro, G. & Zito, R. (1985). Immunological detection of haemoglobin in bones of ancient Roman times and of iron and eneolithic ages. *Proc. Nat. Acad. Sci. 82*:7170-72.

Bender, K. (1984). HLA system. *Bio. Bull. 2*:61-131.

Berry, R.J. (1979). Section I. Genes and skeletons ancient and modern. *J. Hum. Evol. 8*:661-67.

Bodmer, W.F. and Bodmer, J.G. (1978). Evolution and function of the HLA system. In The HLA System. ed
W.F. Bodmer, pp. 309-24. *Brit. Med. Bull. 34*:213-324.

Buikstra, J. (1976). Hopewell in the Lower Illinois Valley. *Northwestern University Archaeology Program Scientific Papers, No. 2.*

Cattaneo, C. (1988). *Palaeoimmunology: the Importance of HLA Antigens in Archaeology and the Feasibility of their Retrieval from Ancient Human Skeletal Material.* Unpublished M.A. thesis, University of Sheffield.

Dausset, J. and Colombani, J. (1973). eds. *Histocompatibility Testing Report.* Copenhagen: Munksgaard.

Davies, K.E. (1986) ed. *Human Genetic Disease: a practical approach.* Oxford: IRL Press.

de Vries, R.R.P., Lai, A., Fat, R.F.M., Nijenhuis, L.E. & Rood, J.J. van (1976). HLA-linked genetic control of host response to *Mycobacterium Leprae. Lancet 2*:1328-30.

Gürtler, L.G., Jager, V., Gruber, W., Hillmar, I., Schobloch, R., Müller, P.K. & Ziegelmeyer, G. (1981). Presence of proteins in human bones 200, 1200, 1500 years of age. *Hum. Biol. 53*:137-50.

Hagelberg, E., Sykes, B., & Hedges, R. (1989). Ancient bone DNA amplified. *Nature 342*:485.

Hart, G.D. (1980). Ancient diseases of the blood. In *Blood Pure and Eloquent* ed. M. Wintrobe, pp. 33-55. New York: McGraw-Hill.

Hedges, R.E.M. and Wallace Carmichael, J.A. (1978). The survival of biochemical information in archaeological bone. *J. Archaeol. Sci. 5*:377-86.

Jones, J.S. (1986). Mummified human DNA cloned. *Nature 314*:576.

Lowenstein, J.M. (1980). Species-specific proteins in fossils. *Naturwiss. 67*:343-46.

Lowenstein, J.M. (1981). Immunological reactions from fossil material. *Philos. Trans. Soc. London B 292*:143-49.

Lowenstein, J.M. (1985) Molecular approaches to the identification of species. *Am. Sci. 73*:541-47.

Loy, T.H. (1983). Prehistoric blood residues: detection on tool surfaces of species origin. *Science 220*:1269-71.

Loy, T.H., Jones, R., Nelson, D.E., Meehan, B., Vogel, J., Southon, J. & Cosgrove, R. (1990). Accelerator radiocarbon dating of human blood in pigments from Late Pleistocene art sites in Australia. *Antiquity 64*:110-16.

Olivier, G., de Castro, M.E. & Almedia (1975). Skull form and differential mortality by tuberculosis. *J. Hum. Evol. 4*:491-95.

Ortner, D.J. and Putschar, W.G.J. (1985). *Identification of Pathological Conditions in Human Skeletal Remains.* Washington: Smithsonian Institute Press.

Paabo, S. (1986). Molecular genetic investigations of ancient human remains. *Cold Spring Harbor Symposia on Quantitative Biology 21*:441-46.

Rothschild, B.M. and Turnbull, W. (1987). Treponemal infection in a Pleistocene bear. *Nature 329*:61-2.

Schiebl, B. (1986). HLA typing. Problems and solutions. *Biotest Mitt. (special edition) 7.* Munich: Biotest AG.

Schwidetzky, I. (1979). Palaeopopulation genetics. *J. Hum. Evol. 8*:669-77.

Stastny, P. (1973). HLA antigens in mummified Pre-Columbian tissues. *Science 183*:864-66.

Steinbock, R.T. (1976). *Paleopathological Diagnosis and Interpretation.* Springfield: Charles C. Thomas.

Svejgaard, A. and Ryder, L.P. (1977). Associations between HLA and disease. In *HLA and Disease* eds. J. Dausset and A. Svejgaard. pp. 46-53. Copenhagen: Munksgaard.

Thompson, J.S. and Thompson, M.W. (1966). *Genetics in Medicine.* Philadelphia: W.B. Saunders.

Thomson, G. (1986). Human HLA genetics and disease associations. In Genetics and Molecular Immunology ed. D.M. Weir *Handbook of Experimental Immunology 3.* pp. 1-102.12 Oxford: Blackwell Scientific Publications.

Ulrich, M.M.W., Perizonius, W.R.K., Spoor, C.F., Sandberg, F. & Vermeer, C. (1987). Extraction of osteocalcin from fossil bones and teeth. *Biochem. Biophys. Res. Comm. 149*:713-19.

Venanzi, F.M. and Rollo, F. (1990). Mummy RNA lasts longer. *Nature 343*:25.

Acknowledgements

I should like to thank Dr. L. Castelletti and Mr. A. Maspero of the Museum and Laboratory of Archaeobiology of Como, Italy, for permission and assistance in using the computer facilities. I am indebted also to Dr. K. Gelsthorpe and Dr. R. Sokol of the Regional Blood Transfusion Centre in Sheffield for their unfailing assistance during the project on the extraction of HLA antigens from ancient human bone.

TABLE 5.1

LIST OF HLA ANTIGENS. (AFTER BENDER, 1984)

HLA-A	HLA-B	HLA-C	HLA-D
A1	B5	CW1	DW1
A2	BW51	CW2	DW2
A3	BW52	CW3	DW3
A9	B7	CW4	DW4
AW23	B8	CW5	DW5
AW24	BW59	CW6	DW6
A10	B12	CW7	DW7
A25	BW44	CW8	DW8
A26	BW45		DW9
AW24	BW59		DW10
A11	B13		DW11
A28	B14		DW12
AW19	B15		
A29	BW62		
AW30	BW63		
AW31	BW16		
AW32	BW38		
AW33	BW39		
AW34	B17		
AW36	BW57		
AW43	BW58		
	B18		
	BW21		
	BW49		
	BW50		
	BW22		
	BW55		
	BW56		
	B27		
	BW35		
	B37		
	B40		
	BW60		
	BW61		
	BW41		
	BW42		
	BW46		
	BW47		
	BW48		
	BW53		
	BW54		
	BW4		
	BW6		

HLA-DR	HLA-DC	HLA-BR	HLA-SB
DRW1	DC1	BR3	SB1
DRW2	DC3	BR4	SB2
DRW3			SB3
DRW4			SB4
DRW5			SB5
DRW6			
DRW7			
DRW8			
DRW9			
DRW10			

TABLE 5.2A

HLA-A, B, C ALLELE FREQUENCIES (%). (AFTER BENDER, 1984)

	Caucasians		Negroes		Orientals	
	Europe	USA	USA	Africa	Japan	USA
HLA-A						
A1	15.8	16.1	8.1	3.9	1.2	2.5
A2	27.0	28.0	16.3	9.4	23.3	43.3
A3	12.6	14.1	7.0	6.4	0.7	0.6
AW23						
A9					37.2	23.2
AW24	8.8	7.3	5.1	2.43		
A25	2.0	2.6	0.4	3.5		
A10					12.7	0.6
A26	3.9	3.4	2.3	4.5		
A11	5.1	5.1	2.8	0.0	6.7	0.0
A28	4.4	4.2	5.8	8.9	0.0	2.8
A29	5.8	3.6	2.3	6.4	0.2	0.6
AW30	3.9	2.9	13.0	22.1	0.5	1.1
AW31	2.3	4.5	2.8	4.2	8.7	19.9
AW32	2.9	3.7	1.9	1.5	0.5	1.1
AW33	0.7	1.2	5.1	1.0	2.0	.06
AW43	0.0	0.0	0.0	0.0	0.0	0.0
Blank	2.2	1.3	16.5	11.0	4.2	1.8
HLA-B						
B5	5.9	5.9	4.9	3.0	20.9	14.0
B7	10.4	10.5	12.6	7.3	7.1	0.6
B8	9.2	10.4	5.5	7.1	0.2	1.7
B12	16.6	13.8	14.0	12.7	6.5	1.7
B13	3.2	2.6	0.4	1.5	0.8	
B14	2.4	5.1	4.6	3.6	0.5	
B18	6.2	3.1	3.6	2.0	0.0	0.6
B27	4.6	5.6	0.8	0.0	0.3	6.2
B15	4.8	5.9	4.7	3.0	9.3	13.7
BW38	2.0	2.5	0.4		1.8	
BW16				1.5		14.5
BW39	3.5	1.4	0.4		4.7	
B17	5.7	4.9	11.2	16.1	0.6	0.0
BW21	2.2	3.8	4.4	1.5	1.5	0.0
BW22	3.6	2.3	3.9	0.0	6.5	0.6
BW35	9.9	8.6	12.5	7.2	9.4	22.1
B37	1.1	1.7	1.2	0.0	0.8	0.0
B40	8.1	9.2	3.9	2.0	21.8	16.6
BW41	0.0	0.0	0.0	1.5	0.0	0.0
BW42	0.0	0.0	0.0	12.3	0.0	0.0
Blank	3.6	2.8	11.0	17.9	7.6	7.8

	Caucasians		Negroes		Orientals	
	Europe	USA	USA	Africa	Japan	USA
HLA-C						
CW1	4.8	3.7	1.9	0.0	11.1	10.1
CW2	5.4	6.0	9.2	11.4	1.4	4.6
CW3	9.4	11.4	8.8	5.5	16.3	16.6
CW4	12.6	10.2	12.9	14.2	4.3	23.4
CW5	8.4	5.2	1.4	1.0	1.2	1.1
CW6	12.6	11.3	0.0	17.7	2.1	0.0
Blank	46.7	52.1	65.8	50.2	53.5	44.2

TABLE 5.2B

SOME HLA-A AND -B ANTIGEN FREQUENCES (%) AMONG EUROPEANS, MIDDLE EASTERNS, INDIANS, MONGOLOIDS. (AFTER DAUSSET AND COLOMBANI, 1973).

HLA	European	Middle Eastern	Indian	Mongoloid
A1	31	24	23	4
A2	48	36	28	33
A3	28	19	14	2
A9	17	34	26	65
A10	12	10	14	14
A11	12	12	33	24
A28	8	12	12	4
A29	8	2	2	2
AW23	4	12	4	4
AW24	14	24	23	63
AW25	4	2	2	4
AW26	8	8	12	10
AW30	4	8	2	4
B5	12	28	34	17
B7	24	6	10	4
B8	21	6	15	2
B12	29	17	19	6
B13	4	6	4	8
B14	8	8	-	-
B27	8	4	4	6
BW35	19	24	23	12

TABLE 5.3

SOME SIGNIFICANT HLA ANTIGEN AND DISEASE ASSOCIATIONS

Disease	HLA	Race	Relative risk
Rheumatoid Arthritis	B27	C	2.0
	DR4	C	3.9
	BW54	O	2.5
	DR4	O	2.8
	DR4	N	5.4
Juvenile Rheumatoid Arthritis	B27	C	3.9
	DR5	C	3.3
Ankylosing Spondylitis	B27	C	69.1
	B27	O	207.9
	B27	N	54.5
Reiter's Disease	B27	C	37.1
Juvenile Diabetes Mellitus	B8	C	2.6
	B15	C	2.2
	DR3	C	3.4
	DR4	C	3.7
	BW54	O	5.7
	DR3	O	4.9
	DR4	O	2.6
	B8	N	2.5
	B15	N	2.2
	DR3	N	3.2
	DR4	N	6.7
Grave's Disease	B8	C	2.6
	DR3	C	3.8
	B35	O	4.4
Celiac Disease	B8	C	7.7
	DR3	O	11.7
	DR7	C	7.7
Psoriasis Vulgaris	B13	C	4.1
	B17	C	5.3
	B37	C	4.0
	CW6	C	7.5
	DR7	C	3.3
	B13	O	3.3
	B17	O	2.0
	B37	O	8.5
	CW6	O	8.5
	DR7	O	7.7
Idiopathic Haemochromatosis	A3	C	6.8
	B7	C	3.0
	B14	C	2.8
Systemic Lupus Erythematosus	B8	C	2.8
	DR3	C	2.7
Goodpasture's Syndrome	DR2	C	13.8
Dermatitis Herpetiformis	B8	O	9.9
	DR3	C	17.3
Multiple Sclerosis	B7	C	1.9
	DR2	C	2.8
Myasthenia Gravis	B8	C	3.4
Narcolepsy	DR2	O	358.1
	DR2	C	129.8

C = Caucasian O = Oriental N = Negro

The Relative Risk Factor is an indication of how much more often the carrier of an antigon will be affected by the disease, i.e. the risk of developing the disease within the carrier of a specific antigen relative to that for individuals not carrying that antigen (Bernal, 1986). A relative risk greater than 1 is present when an antigen is more frequent in the patients than in the control (Svejgaard and Ryder, 1977).

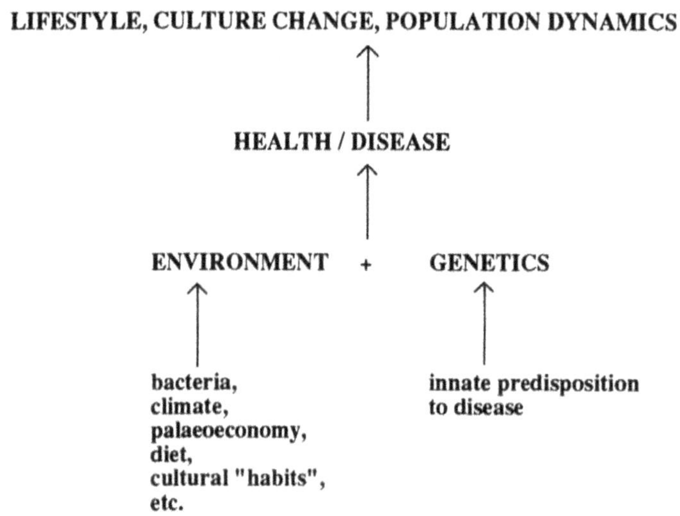

Figure 5.1 Scheme showing the introduction of genetic (along with environmental) information in the reconstruction of the health status and therefore of the lifestyle and dynamics of a skeletal population.

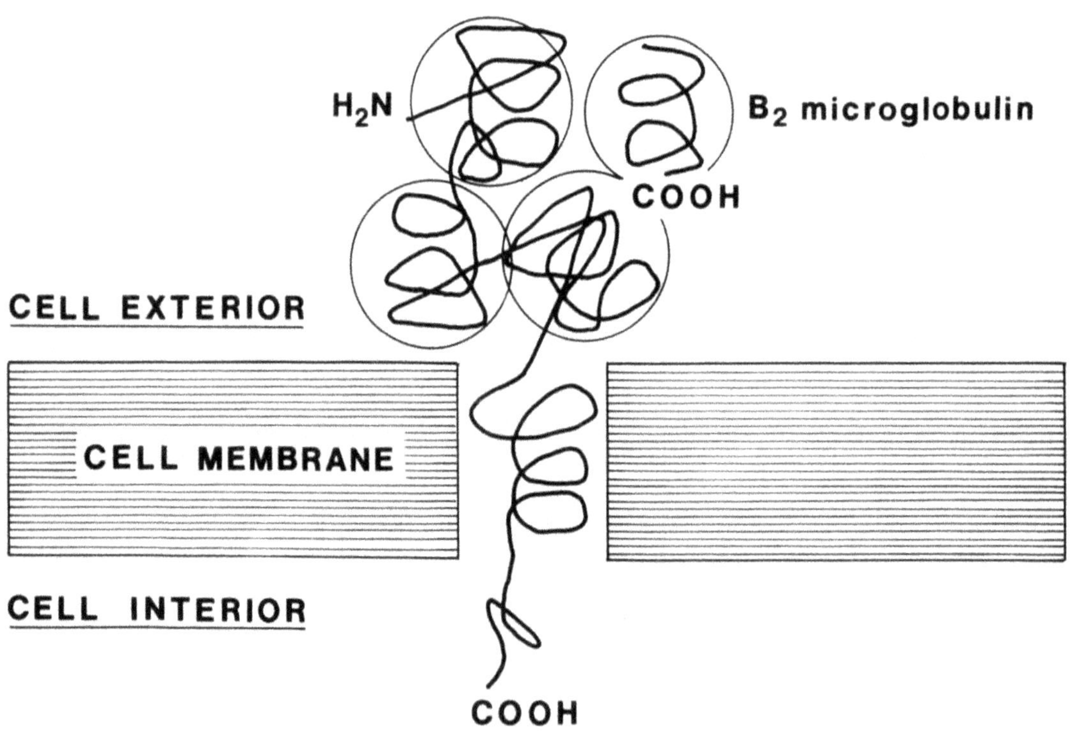

Figure 5.2 Drawing of a class I HLA antigen. Class I HLA antigens have a molecular weight of 45000. Each chain is inserted into the cell membrane so that a short segment is within the cell, one segment across the lipid bilayer of the cell membrane, and the largest segment is exposed to the external environment and is folded into three domains. Class I antigens make up circa 1% of the membrane proteins. Associated to this bigger segment is a small protein, microglobulin. (After Alberts *et al.*, 1983).

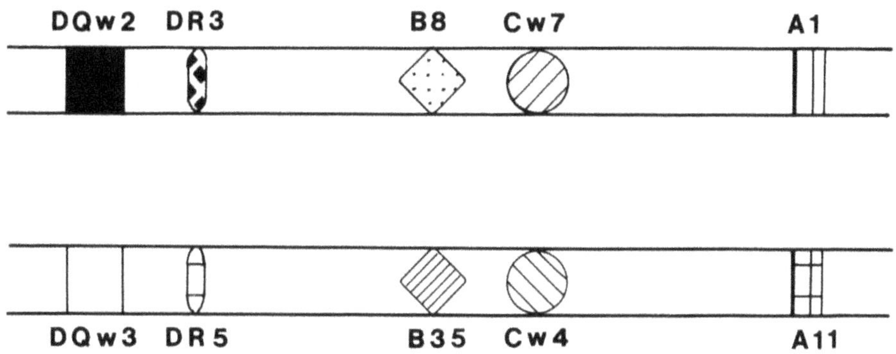

Figure 5.3 Schematic example of two haplotypes in an individual. Each houses the A,B,C,DQ,DR loci, but the alleles may be of different "types", e.g. Al and All. (After Schiebl, 1986).

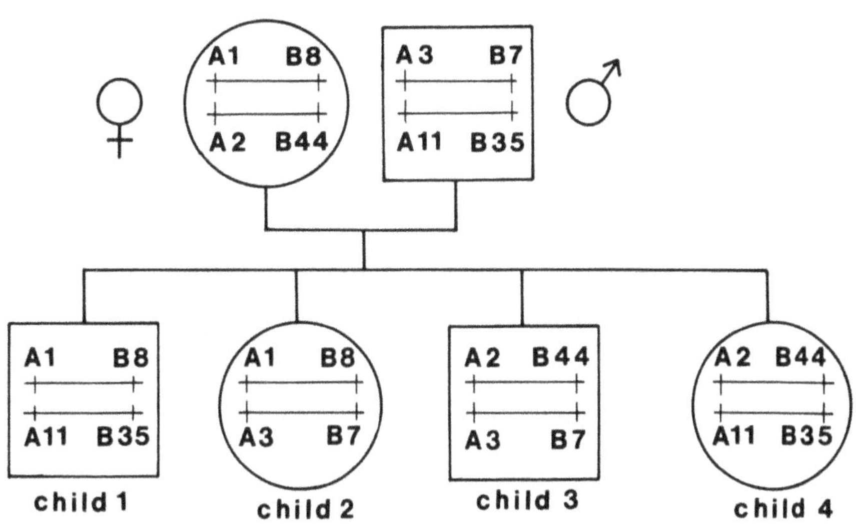

Figure 5.4 Inheritance units in a family segregtion analysis. The maternal and paternal haplotypes are shown. These 4 haplotypes can recombine to give four possible combinations in offspring. Each haplotype is inherited as a unit. (After Bender, 1984).

A MORE ACCURATE RECONSTRUCTION OF THE CULTURE AND DYNAMICS OF ANCIENT POPULATIONS

↑

+ other environmental and archaeological data

↑

SOUNDER BASE FOR THE RECONSTRUCTION OF THE HEALTH STATUS

↑

GENETIC (INNATE) SUSCEPTIBILITY TO DISEASE

↑

EXTRACTION OF GENETICALLY INFORMATIVE MOLECULES FROM SKELETAL MATERIAL

Figure 5.5 General scheme showing the use of genetically informative molecules, extracted from skeletal remains, in the reconstruction of the dynamics and culture of a population.

6. DENTAL HISTOLOGY AS AN INDEX OF PAST COMMUNITY HEALTH

Simon Hillson

Community Health

The concept of health might be summarised in terms of a community's dietary status, the incidence of infectious and other disease, or social conditions which may impinge on both mental and physical well being. It is very difficult to define a general index of health, even in living human communities. Most major studies use the growth rate of children in the community as at least a general indicator. Growth requires a surplus in the body's resources, over and above what is required for day-to-day maintenance. Growth rate is therefore limited by dietary deficiencies (which fail to supply the required surplus) and diseases (which cause the body to use up its surpluses in fighting them). Even emotional stress may decrease growth rate (Harrison *et al*, 1977). With living communities, estimates of growth rate are calculated in various ways and then compared with growth rate in a reference community which is supplied with an abundant diet and has good health care - usually in the USA or Europe. This reference data is accumulated over a long period by studying children when they visit clinics at regular intervals. Individual children are followed through their lives, measuring stature, head circumference, weight and a host of other general indices of overall body size. Large numbers of children supply the data for charts which summarise the variation in size achieved at particular ages in terms of percentiles. Charts like these are frequently issued to parents so that they can plot their own child's progress and see if he or she lags seriously behind. When the health of another community is to be assessed, for example in a Third World country, it is usually not possible to follow individual children over a period of years. A much shorter timescale is needed and so growth rate is established by an alternative procedure. Large samples of children are measured and their ages are determined by asking their parents, examining medical records etc. Age can then be plotted against size in a way that may be compared with the percentiles of the reference charts. There are clearly large possibilities for error in this approach, but it has been used with success for many years.

In an ideal world, it would be possible to apply similar methods to archaeology, but any attempt to do this instantly involves a circular argument. It is possible to take measurements of the skeleton which reflect overall body size - for example, lengths of long bones in the limbs. It is also possible to obtain a reasonable estimate of age-at-death in children and potentially the measurements could be plotted against these ages. The difficulty is that all the age estimation methods commonly used on the remains of children are themselves based on phenomena related to the growth process. Any reliability that they have as age estimators relies, in fact, on the assumption there is *no difference* in growth rate between a once-living group of people represented by archaeological material and the modern reference community. If growth rate is to be studied in archaeology, then some independent measure of age is required.

Layering in Teeth

Independent clocks do exist in human physiology. Many diurnal and other rhythms relating to the growth process have been discovered (Dean, 1987). Only the hard tissues of the body survive on most sites and the best examples of growth rhythms amongst these occur in the tissues of the teeth. The enamel that forms the outer jacket of the tooth crown, and the dentine which makes up the main body of the tooth and root, both bear records of daily rhythms and of rhythms with longer periodicities in the form of microscopic layered structures.

a) Dentine layers

Thin sections of dentine, viewed under the polarising microscope, frequently show a fine, regular pattern of lines (Figure 6.1). These are usually known as the *incremental lines of von Ebner*, after the author of one of the classic papers describing them, but they may also be described as *Andresen's lines* (Kawasaki *et al*, 1980) or *Mummery's lines* (Schmidt and Keil, 1971). Experimental studies with pigs (Yilmaz *et al*, 1977), using regular injections of marker dyes at two-week intervals, have suggested that these lines represent a daily (24 hour) layering of dentine apposition. The lines vary both in spacing and prominence and in some individuals there may also be lines of marked distortion to the structure of the dentine matrix, spaced at intervals in the layering sequence. These are usually known as the *contour lines of Owen*, after the great anatomist Sir Richard Owen who first described them, but there is some confusion over the exact meaning of this term and one author at least prefers to call them *Schreger lines*

(Bradford, 1967). The prominence of both what I shall call von Ebner's lines and the contour lines of Owen is under some kind of systemic control: in one individual, patterns of lines can be matched between all teeth which were being formed at the same time. It is therefore possible to match corresponding teeth from the right and left sides of the jaw, and from upper and lower jaws. It is also possible to match teeth of different class whose dentine formation time happends to overlap (Figure 6.2). In this way it may be possible to match, for example, a canine with a first molar. The first practical application of this phenomenon was by the Swedish forensic dentist Gösta Gustafson (1947) who, in a blind test, was able to match up the teeth of three different individuals by these means. The pattern of dentine layering is highly individual and clearly has potential as a means of identification in forensic studies. Gustafson did not suggest counting the lines of von Ebner, but this would be a logical extension of his work. Matching of different teeth in the sequence could provide an extended chronology of dentine formation, very much in the same way that tree ring sequences do. Such a chronology could even be given an accurate starting point, because the point of birth is clearly marked by a prominent line (called the *neonatal line*) in the dentine of teeth which were forming at the time (Schour, 1936). The neonatal line can be seen in all deciduous teeth and in a large proportion of permanent first molars. By counting layers from the neonatal line, it would theoretically be possible to determine precisely when each tooth in the dentition started and finished. If teeth were still being formed at the time of death, the last layer formed would yield an exceptionally precise estimate of age-at-death but, more importantly from the point of view of growth studies, the chronology of tooth formation would act as an index of growth rate which could be compared with radiographic studies of dental growth rate in living populations (e.g. Moorrees *et al*, 1963; Anderson *et al*, 1976).

There are, however, considerable difficulties even at a theoretical level. Different authors mean different things by the terms they use to describe the lines. Estimates of spacing vary greatly and the evidence for the daily rhythm underlying the structures seen in the microscope is not always clear. Part of the trouble is that the formation of dentine involves two separate phases, and it may be difficult to determine which is responsible for a given structure or line in the fully formed tissue. The first phase of formation is the deposition of an organic matrix called *predentine*. This consists largely of felted mats of the fibrous protein *collagen*, together with other organic material known as ground substance. The second phase is the mineralisation of the tissue. To some extent the tiny crystals of mineral follow the collagen fibrils, but the dominant pattern of mineralisation is *calcospheritic*. That is, the crystals grow out radially from individual centres to form spherical bodies which grow until they meet and fuse with one another (Figure 6.1). This pattern occurs independently to the collagen fibrils and other structures of the original dentine matrix but, in fresh dentine, organic structures are still preserved inside the fully mineralised tissue. Even if all organic matter is removed chemically from fresh dentine (an "anorganic" preparation), the overall arrangement of organic components may still be visible under the microscope because mineralisation has to some extent followed their structure, in addition to following the dominant calcospheritic mineralisation (Schmidt and Keil, 1971; Jones and Boyde, 1984). Features of the original predentine matrix may be visible due to this phenomenon even in archaeological specimens which have long since lost most of their collagen.

The incremental lines of dentine represent both the process of matrix deposition and mineralisation. They may be structures which were originally formed in the predentine organic matrix, and then partially emphasised by the pattern of mineralisation that followed them. They may also represent deficiencies of mineralisation which occurred independently to the predentine layering. This situation is most clearly apparent as bands of *interglobular spaces* (Figure 6.1), where the calcospherites have failed to coalesce, leaving a zone of poorly mineralised predentine between. These have been related experimentally to growth disturbing factors, such as fever in rabbits (Bermann *et al*, 1939) and rats (Garrison, 1940), and deficiencies of vitamins A and D in dogs (Mellanby, 1929). As the front of mineralisation lags some 15μ behind the front of matrix formation (Mjor, 1984) and does not always follow the same line, the position of a band of interglobular spaces does not coincide with that of the predentine layer which was formed at the same time. The contour lines of Owen (above), as originally described, are features of the original predentine matrix (Bradford, 1967). Some texts (Schmidt and Keil, 1971) insist instead that they are lines of interglobular spaces, others equate them with less marked lines of poor mineralisation (Bhaskar, 1986) and still others (ten Cate, 1980) report that the term is commonly used to mean exaggerated lines of von Ebner. These last are in an even greater state of turmoil. Careful work by Schmidt and Keil (1971) has shown that von Ebner's lines are part of the organic matrix and are therefore formed during the first phase of dentine production. The proof for this is that, when the mineral component of fresh dentine is removed chemically (a "demineralised" preparation), the lines are still clearly visible. They are, however, also visible (rather more faintly) when the organic component is removed, so the mineralisation must, to some extent, follow the organisation of the lines in the predentine matrix. What causes the layering in the predentine is not clear. Some authors (Schmidt and Keil, 1971) suggest that it is due to regular alternations in the direction of collagen fibrils, whilst others (Jones and Boyde, 1984) suggest that it may represent cyclical variation in the proportion of ground substance to collagen.

In addition, it is not entirely clear which lines should be identified as "lines of von Ebner". Most textbooks give a similar range of figures for the spacing of the lines: about 5μ (Osborn, 1981), 3.5μ to 6μ depending on the position in the tooth (ten Cate, 1980), 4 to 8μ (Bhaskar, 1986). Bradford (1967) defined two types of regular incremental line, one on average 25μ apart and the other about 3-4μ apart. Kawasaki *et al* (1980) supported this view when they found more prominent lines at 20μ intervals in the dentine under the tooth crown, with four fainter and more intermittent lines in between them spaced at 4μ intervals. They suggested that the original publications of von Ebner and Andresen were both describing the coarser interval of lines only. So what should be called "lines of von Ebner"? In my own sections, both of freshly extracted teeth and archaeological specimens, the most prominent regular patterns of incremental lines have a spacing of 18 to 28μ in the dentine immediately under the centre of the tooth crown. Careful examination of the dentine between major lines in my sections does reveal indications of lines between 3.5μ and 5μ apart. The major lines do, however, become more closely spaced as they are followed down towards the sides of the crown.

The situation is further confused by attempts to link the spacing of the lines to estimates for the rate of formation of dentine. The lines that Yilmaz *et al* (1977, see above) related to a diurnal rhythm in pigs appear to be about 6.25μ apart when scaled from their published microphotograph. Other experiments based upon the injection of marker dyes into experimental animals have produced estimates for dentine apposition rate ranging from 16μ per 24 hours in rats, to 4μ in rhesus monkeys (Symons, 1967). Some drugs given therapeutically in man also produce a marker line in the dentine and these have also yielded estimates of around 4μ per 24 hours in the deciduous teeth of children (Schour and Poncher, 1937). Kawasaki *et al* (1980) used a different technique, comparing the dentine layers with incremental structures in the neighbouring enamel which are thought to represent daily growth (see below). They also suggested that, under the central part of the crown, dentine grows at a rate of 4μ per day. On the basis of this evidence, it seems likely that it is the finer incremental lines of the dentine which represent a 24 hour growth rhythm.

In spite of the considerable potential of dentine layering, a great deal of basic histological research is needed before this technique is likely to be a practical proposition, even in fresh, well-preserved material.

b) Enamel layers

Dental enamel is at present a more realistic proposition because, being almost wholly mineral, it is usually much better preserved than dentine in archaeological contexts. Enamel layering is also rather clearer, more consistently visible and presents fewer difficulties of interpretation. Under the scanning electron microscope, fractured surfaces which have been etched for a few seconds in acid show the major features of enamel structure well. The most prominent features are bundles of tiny crystals, about 5μ across and called *prisms*. The prisms alternately swell and constrict along their length, with a regular interval of around 4μ (Figure 6.3). This feature can also be seen in thin sections of enamel, examined under the light microscope, as alternate light and dark markings along the prisms. They are known as *prism cross striations* and there is now little doubt that they reprsent a daily rhythm (Dean, 1987). Counts of cross striations between prominent markers in the enamel have been matched with similar counts of the lines of von Ebner between corresponding markers in dentine (Kawasaki *et al*, 1980).

Enamel also shows a roughly weekly rhythm. This is most clearly and consistently seen in thin sections of enamel under the the optical microscope and usually appears as dark, fuzzy lines cutting diagonally across the boundaries of the prisms (Figure 6.4). These are named *brown striae of Retzius*, after the Swedish anatomist, Anders Retzius, who published the classic description (1837). Cross striations may be counted between neighbouring brown striae of Retzius and the count varies between seven and nine, with an average in most studies of about eight (Dean, 1987). The variation in counts may well be due to experimental error, because brown striae are rarely clear-cut features. They tend to become indistinct under higher power magnifications and in thinner sections, whereas cross striations become less distinct in thicker sections and at lower magnifications. It therefore seems reasonable to postulate a roughly eight cross striation, and therefore eight day, rhythm to the formation of the layers which are marked out by the brown striae of Retzius.

The brown striae vary markedly in form. Some are more prominent than others. In some cases they have a sharp line at their centre which appears to link up a line of cross striations whilst in other cases, they may be very diffuse features. This variation seems to be again under some form of systemic control, because groups of prominent brown striae may be matched between teeth in one individual, and the counts of cross striations between them may be matched also. This was demonstrated by Fujita (1939) and was investigated further by Anna-Greta Gustafson (1955) as a means of matching teeth from one individual for forensic purposes. As in dentine, the period of enamel formation (Figure 6.5) varies between teeth and the different periods may be arranged in an overlapping series from birth up to around 15 years of age. Starting from the neonatal line, which is often visible in the permanent first molar, Boyde (1963a; 1963b) was able to use counts of cross striations and matching of brown striae to arrive at exceptionally precise age estimates for individuals who had died whilst their teeth were still forming. Bullion (1987) was able, in a number of cases, to produce sequences of 350 or more brown striae which represented seven years of tooth growth in childhood. It is possible to estimate crown formation timing from these sequences and, as in dentine, this could be used as a very precise index of growth rate which could be compared between different ancient communities and with modern reference material. It is possible to construct such sequences routinely because of the good preservation of enamel and the clarity of brown striae in most sections. Since counts of cross striations in enamel may be matched with counts of the lines of von Ebner in dentine (Kawasaki *et al*, 1980, see above) it may be possible to combine enamel and dentine sequences to cover patches where either tissue is poorly preserved or the layering lacks clarity.

In spite of their potential, such layer sequences are rarely attempted even in enamel. The main reason for this must be the time-consuming and destructive sectioning which is needed. A full sequence would probably require first, second and third molars, a premolar, canine and incisor all to be sectioned. Each needs to be encapsulated into a resin block (Hillson, 1986) because ancient teeth are usually brittle and shatter when sectioning is attempted. The best impregnation procedures take several weeks and sectioning, several hours. In addition to this, the teeth are destroyed largely, although high quality replicas can be made before sectioning. For all these reasons, studies of enamel layering are focussing primarily on the surface of the tooth crown.

The brown striae of Retzius mark out the pattern of layering in the crown (Hillson, 1986). The first enamel layers are laid down under the cusps or highest ridges of the crown as a series of domes, one on top of another. These domes increase in area and overlap one another down the sides of the crown. When the full height of the cusps or ridges has been attained, the pattern changes to sleeve-like layers, laid down on top of the domes and again overlapping one another in a sequence down the crown sides. In this way, the first months of enamel formation are hidden from the surface under the last dome-shaped layer to be formed. Thereafter, each layer rises at an angle to the surface and leaves its mark there. Where each brown stria meets the surface, a furrow is formed, running around the circumference of the crown (Figures 4 and 6). The edge of the layer in between brown striae is represented as a ridge between the furrows. A whole series of these can be seen in unworn teeth, extending down the full height of the crown and looking like a series of waves, or overlapping strata. The ridges and furrows vary greatly in form and have a variety of names (Risnes, 1984), but are together normally called *perikymata*. This was the name given to them by Gustav Preiswerk in 1895. Each brown stria is thought to be represented by the furrow of a perikyma and each furrow is thought to extend fully around the circumference of the crown (Risnes, 1985a; 1985b). They may therefore be counted in the same way as brown striae and, assuming the same eight day average growth rhythm, may be used to estimate the time taken for the crown surface to form.

Estimates of crown formation time established by these means have been used to arrive at improved estimates for the age at death of fossil hominids (Bromage and Dean, 1985; Dean *et al*, 1986). Further potential lies in the possibility of building up extended layer sequences, as with the lines of von Ebner and the brown striae of Retzius. The perikymata vary in prominence, regularity and spacing. Part of this is related to the way in which the crowns grow: the grooves of the perikymata are further apart near the occlusal tip of the crown than they are near the cervix (where the crown meets the root). This regular progression is seen in all teeth, but there are also detailed variations which are not part of it. Some perikymata may be particularly prominent and several of these may form a pattern whichcan be matched between different teeth from one individual. Bands of irregular perikymata may also be matched between teeth and this suggests that both are related to systemic factors in the same way that prominence of brown striae and lines of von Ebner are. Perikymata are also related to macroscopically visible defects of the crown surface. These take various forms, but are most commonly marked grooves or bands of pitting which run round the circumference of the crown and are large enough tobe seen by the naked eye. Such defects have long been observed by dental surgeons and one of the key early studies was by Robert Bunon in 1746. Bunon called the defects *erosions* and demonstrated them in the unerupted teeth of children who had suffered from such diseases as rickets, scurvy, measles and smallpox. Defects could have occurred in these teeth only by interruptions to the growth process. The link with deficiencies in vitamins A and D was established by Lady Mellanby in her experiments with beagle dogs (1929) and a variety of clinical studies in man have since confirmed a connection with various nutritional deficiencies, infectious diseases and other factors (Pindborg, 1982). The defects are nowadays referred to as *enamel hypoplasia*, meaning under-development of the enamel. In some cases they can be matched with interglobular spaces in dentine (above), but this is by no means always so. Even though their general relationship to growth disrupting factors such as fever and dietary deficiency seems clear, it is difficult to link particular defects to particular disturbances. Sarnat and Schour (1941; 1942) matched only 50% of defects with episodes of fever in the medical records of 60 children. In an unpublished study of 28 young men, I matched 70% of defects with episodes recorded in their medical records only when the most minor defects were included.

In spite of the difficulties in explaining individual defects in living people, they are clearly of systemic origin and can be matched, along with irregularities of perikymata, between teeth in one individual. On closer study, the groove-like defects are themselves seen to be formed by variations in the spacing of perikymata (Figure 6.6; Hillson and Jones, 1989). The perikymata/hypoplasia sequences of different teeth may therefore be matched up to produce an extended sequence of the kind suggested for lines of von Ebner and brown striae of Retzius. This was suggested by myself (Hillson, 1979; 1986) and has since been tested by one of my research students (Bullion, 1987), who was able to construct composite sequences of up to 200 perikymata, representing around four and a half years of growth. These sequences may be linked approximately to chronological age if the average ages for the start and completion of the hidden dome-like layers are known for some of the teeth in the sequence. Bullion (1987) has made a pioneering attempt at establishing these for a range of teeth, by counting brown striae of Retzius in tooth sections. Perikymata can be difficult to study and are quite rapidly lost by abrasion of the crowns in adult teeth. The potential, however, is great because both perikymata and hypoplasia can be examined non-destructively in casts of the crown surface. These can be made from the teeth of living people as well as ancient burials. They may yield the same kind of information on dental growth rate and growth disturbances as the brown striae of Retzius and the incremental lines of dentine, so that they offer the best hope of a growth index that could be determined routinely and compared with living communities of people.

Conclusion

The layered structure of enamel and dentine offers what is probably the only possibility in archaeology for studying growth-rate in relation to an independent measure of time. These two tissues are likely to prove a valuable resource in both archaeology and anthropology, but realising their potential still requires a great deal of hard work on basic histology and physiology.

References

Anderson, D.L., Thompson, G.W. & Popovitch, F. (1976). Age of attainment of mineralisation stages of the permanent dentition. *J. For. Sci. 21*:191-200.

Bermann, M., Edwards, L. & Kitchin, P. (1939). Effects of artificially induced hyperpyrexia on tooth structure in the rabbit. *Proc. Soc. Exp. Biol. Med. 41*:113-15.

Bhaskar, S.N. (1986). ed. *Orban's Oral Histology and Embryology*. St. Louis: C.V. Mosby.

Boyde, A. (1963a). Estimation of age at death of young human skeletal remains from incremental lines in the dental enamel. *3rd International Meeting in Forensic Immunology, Medicine, Pathology & Toxicology (London April 16th-24th 1963) Plenary Session IIA.*

Boyde, A. (1963b). Unpublished data quoted by Miles, A.E.W. In *Dental Anthropology* ed. D.R. Brothwell. pp. 191-210. London: Pergamon Press.

Bradford, E.W. (1967). Microanatomy and histochemistry of dentine. In *Structural and Chemical Organization of Teeth*. ed. A.E.W. Miles. pp. 3-32. Orlando: Academic Press.

Bromage, T.G. and Dean, M.C. (1985). Re-evaluation of the age at death of immature fossil hominids. *Nature 317*:525-27.

Brothwell, D.R. (1963). ed. *Dental Anthropology*. London: Pergamon Press.

Bullion, S.K. (1987). *Incremental Structures of Enamel and their Applications to Archaeology*. Unpublished Ph.D. thesis, University of Lancaster.

Dean, M.C. (1987). Growth layers and incremental markings in hard tissues: a review of the literature and some preliminary observations about enamel structure in *Paranthropus boisei*. *J. Hum. Evol. 16*:157-72.

Dean, M.C., Stringer, C.B. & Bromage, T.G. (1986). A new age at death for the Neanderthal child from Devil's Tower, Gibraltar and the implications for studies of general growth and development in Neanderthals. *Am. J. Phys. Anthropol. 70*:301-09.

Fujita, T. (1939). Neue feststellungen über die Retzius'shen parallelstreifen des Zahnschmelzes. *Anat. Anz. 87*:350-55.

Garrison, L. (1940). The effect of fever on the development of the rat incisor. *J. Dent. Res. 19*:215-25.

Gustafson, A-G. (1955). The similarity between contralateral pairs of teeth. *Odontol. Tidskr. 63*:245-48.

Gustafson, G. (1947). Microscopic examination of teeth as a means of identification in forensic medicine. *J. Am. Dent. Assoc. 35*:720-24.

Gustafson, G. and Koch, G. (1974). Age estimation up to 16 years of age based on dental development. *Odontol. Revy 25*:297-306.

Harrison, G.A., Weiner, J.S., Tanner, J.M. & Barnicot, N.A. (1977). *Human Biology* 2nd edition. Oxford: Oxford University Press.

Hillson, S.W. (1979). *Human Biological Variation in the Nile Valley, in Relation to Environmental Factors*. Unpublished Ph.D. thesis, University of London.

Hillson, S.W. (1986). *Teeth*. Cambridge: Cambridge University Press.

Hillson, S.W. and Jones, B.K. (1989). Instruments for measuring surface profiles: an application in the study of ancient human tooth crown surfaces. *J. Archaeol. Sci. 16*:95-105.

Jones, S.J. and Boyde, A. (1984). Ultrastructure of dentin and dentinogenesis. *Dentin and Dentinogenesis.* ed. A. Linde. pp. 81-134. Boca Raton: CRC Press Inc.

Kawasaki, K., Tanaka, S. & Ishikawa, T. (1980). On the daily incremental lines in human dentine. *Arch. Oral Biol. 24*:939-43.

Linde, A. (1984). ed. *Dentin and Dentinogenesis.* Boca Raton: CRC Press Inc.

Mellanby, M. (1929). Diet and teeth: an experimental study. Part I. Dental structure in dogs. *Medical Research Council, Special Report Series Number 140.* London: HMSO.

Miles, A.E.W. (1963). Dentition in the assessment of individual age in skeletal material. In *Dental Anthropology* ed. D.R. Brothwell. pp. 191-210. London: Pergamon Press.

Miles, A.E.W. (1967). ed. *Structural and Chemical Organization of Teeth.* Orlando: Academic Press.

Mjor, I.A. (1984). The morphology of dentin and dentinogenesis. In *Dentin and Dentinogenesis.* Vol. 1. ed. A. Line. pp. 1-18. Boca Raton: CRC Press Inc.

Moorrees, C.F.A., Fanning, E.A. & Hunt, E.E. (1963). Age variation of formation stages for ten permanent teeth. *J. Dent. Res. 42*:1490-1502.

Osborn, J.W. (1981). ed. *Dental Anatomy and Embryology: A Companion to Dental Studies 1(2).* Oxford: Blackwell Scientific Publications.

Pindborg, J.J. (1982). Aetiology of developmental enamel defects not related to fluorosis. *Int. Dent. J. 32*:123-34.

Retzius, A. (1837). Bemerkungen über den inneren Bau der Zäthne. *Archiv. Anat. Physiol.* 486-566.

Risnes, S. (1984). Rationale for consistency in the use of enamel surface terms: perikymata and imbrications. *Scand. J. Dent. Res. 92*:1-5.

Risnes, S. (1985a). A scanning electron microscope study of the three-dimensional extent of Retzius lines in human dental enamel. *Scand. J. Dent. Res. 93*:145-52.

Risnes, S. (1985b). Circumferential continuity of perikymata in human dental enamel investigated by scanning electron microscopy. *Scand. J. Dent. Res. 93*:185-91.

Sarnat, B.G. and Schour, I. (1941). Enamel hypoplasia (chronologic enamel aplasia) in relation to systemic disease: a chronologic, morphologic and etiologic classification. *J. Am. Dent. Assoc. 28*:1989-2000.

Sarnat, B.G. and Schour, I. (1942). Enamel hypoplasia (chronologic enamel aplasia) in relation to systemic disease: a chronologic, morphologic and etiologic classification. *J. Am. Dent. Assoc. 29*:67-75.

Schmidt, W.J. and Keil, A. (1971). *Polarization Microscopy of Dental Tissues.* Oxford: Pergamon Press.

Schour, I. (1936). Neonatal line in enamel and dentin of human deciduous teeth and first permanent molar. *J. Am. Dent. Assoc. 23*:1946-1955.

Schour, I. and Poncher, H.G. (1937). The rate of apposition of human enamel and dentin as measured by the effects of acute fluorosis. *Am. J. Dis. Child. 54*:757-76.

Symons, N.B.B. (1967). Dentinogenesis. In *Structural and Chemical Organization of Teeth.* ed. A.E.W. Miles. pp. 285-324. London: Academic Press.

ten Cate, A.R. (1980). ed. *Oral Histology.* St. Louis: C.V. Mosby.

Yilmaz, S., Newman, H.N. & Poole, D.F.G. (1977). Diurnal periodicity of von Ebner growth lines in pig dentine. *Arch. Oral Biol. 22*:511-13.

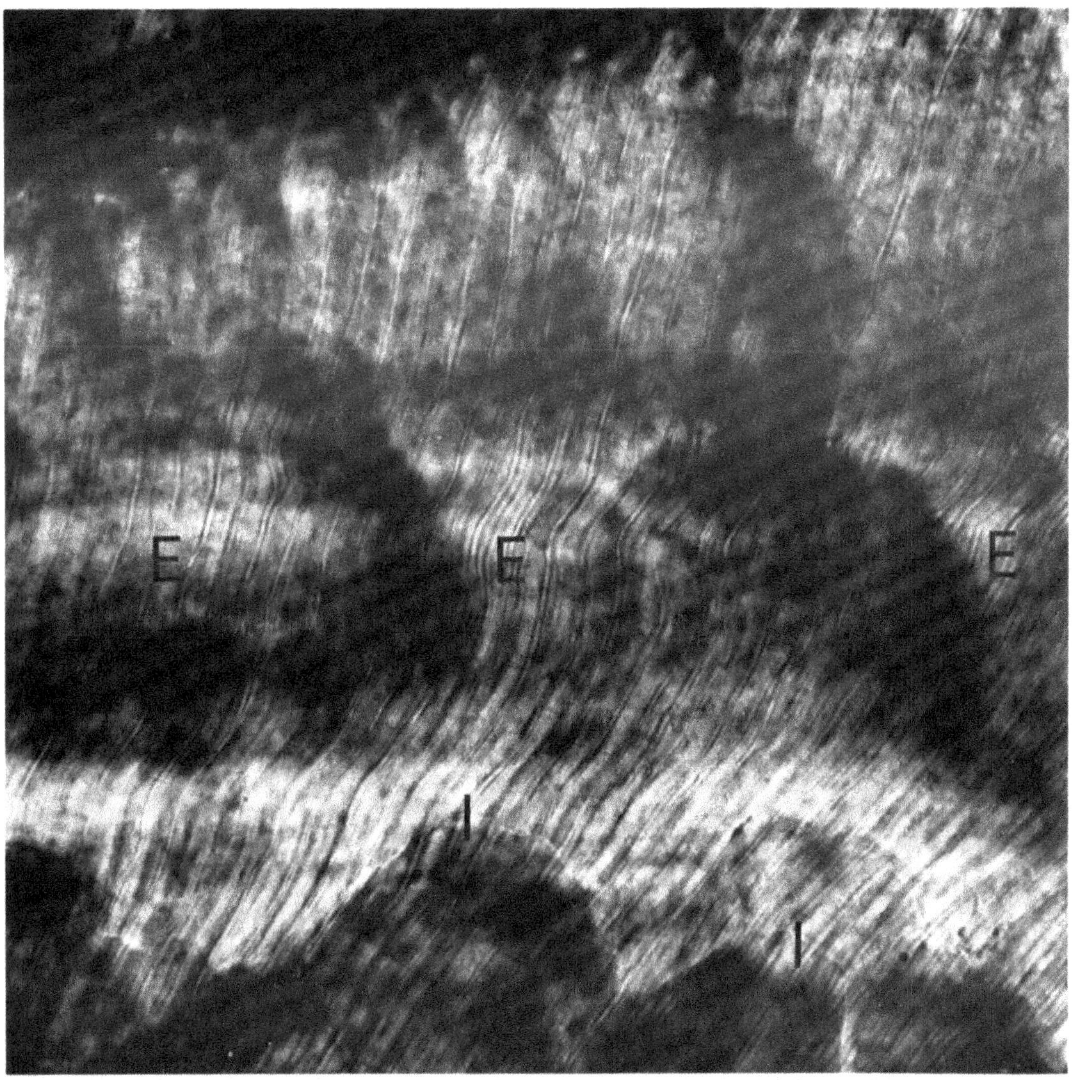

Figure 6.1 Incremental structures in the dentine of a human canine from the site of Kerma, Nubia (c. 1720-1550 B.C.). Material from the Duckworth Laboratory, University of Cambridge. Von Ebner's lines are the fuzzy dark bands running horizontally across the field of view (one is marked E-E-E). The calcospherites can be seen as dark, cloud-like masses. Towards the bottom of the field of view, some are outlined by narrow patches of interglobular dentine (Marked I). The sharp, sinuous lines running up and down the photograph are the dentinal tubules, not an incremental structure, but a major feature of dentine histology. Polished thin section viewed by polarising microscopy. Field width c. 50μ.

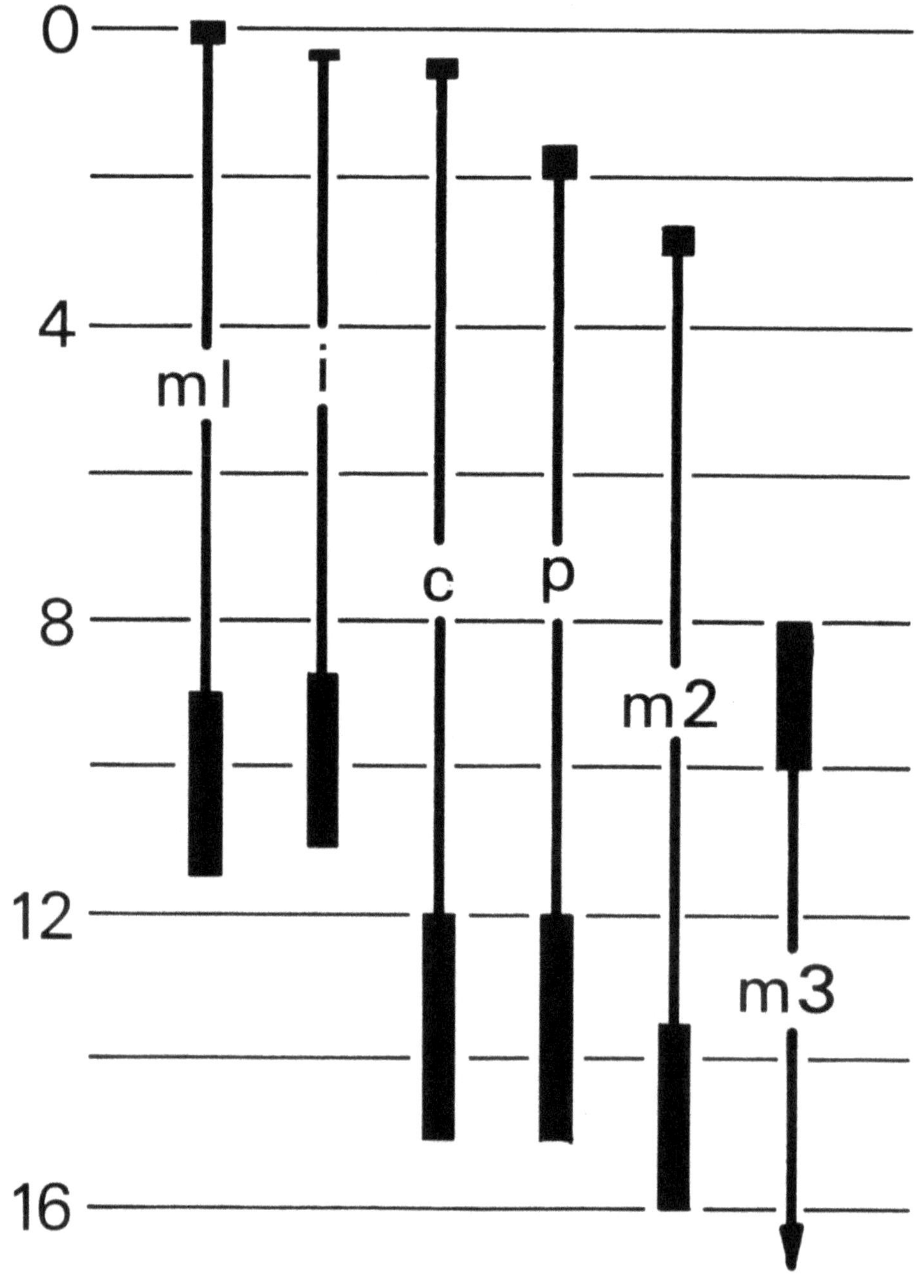

Figure 6.2 Ranges of dentine formation times for various human permanent teeth (data from Gustafson and Koch, 1974). Codes for teeth: ml = first molar; i = incisors, c = canine, p = premolars, m2 = second molar, m3 = third molar. Scale in years after birth.

Figure 6.3 Prisms in the enamel of a human first molar from the Coppergate site, York. The prisms run up and down diagonally and show the alternate swellings and constrictions (prominent in the area marked C) along their length which are thought to correspond to cross striations. Fractured, etched specimen, examined by scanning electron microscopy, using secondary electron mode. Field width c. 60μ.

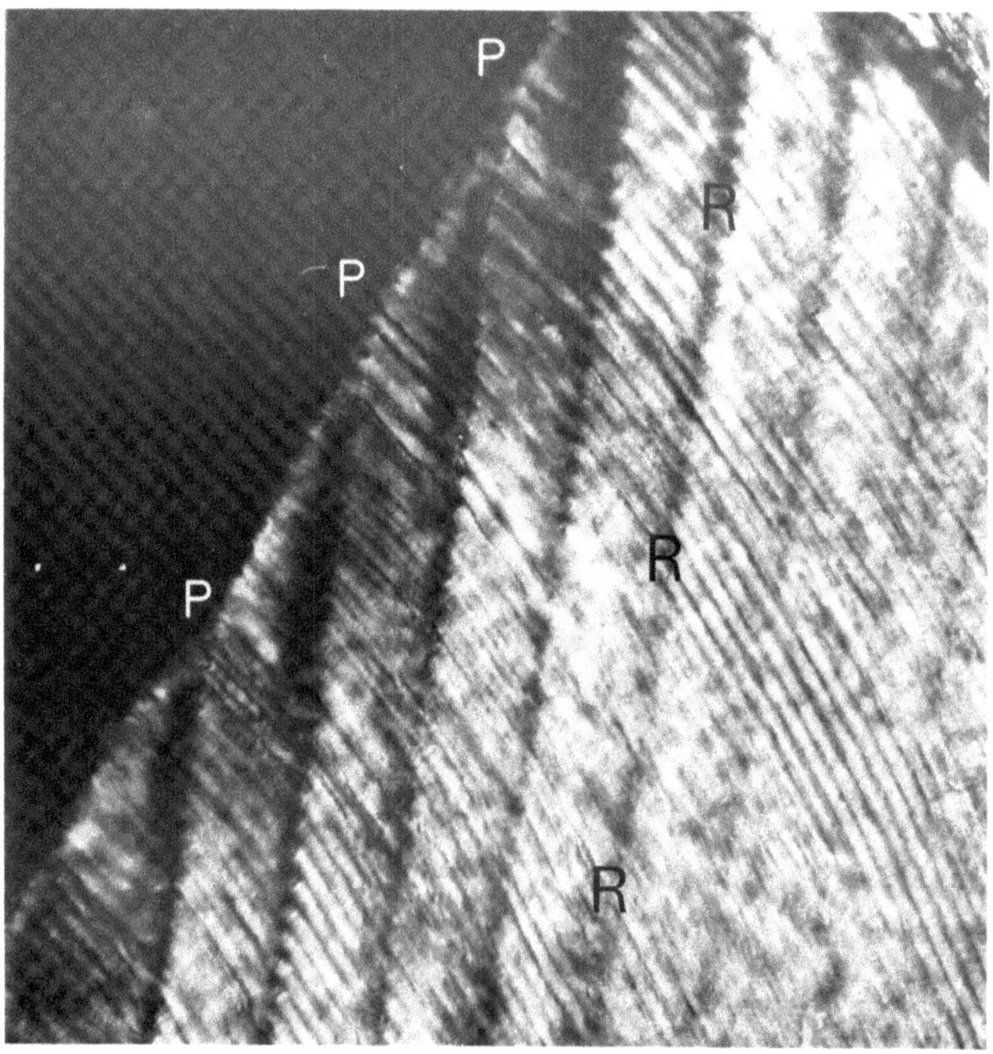

Figure 6.4 Brown striae of Retzius in the enamel of a human canine from Kerma, Nubia (c. 1750-1550 B.C.). Material from the Duckworth Laboratory, University of Cambridge. The striae (one is marked R-R-R) are running almost vertically up the photograph, to meet the surface of the crown where slight indentations marking the furrows of the perikymata (marked P) can be seen. The prisms can be seen cutting across the line of the brown striae diagonally. Polished thin section examined by polarising microscopy. Field width c. 250μ.

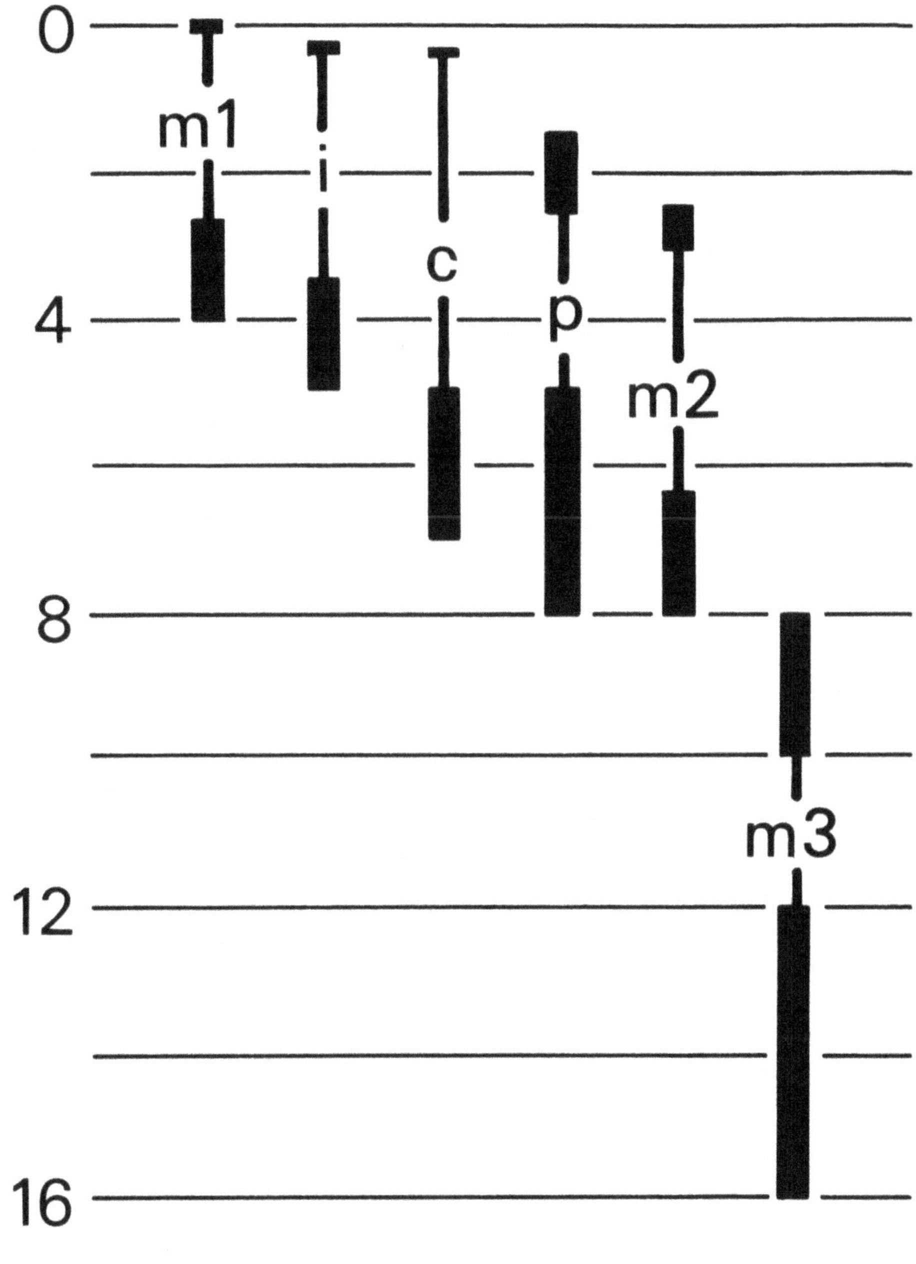

Figure 6.5 Ranges of enamel formation times for various human permanent teeth (data from Gustafson and Koch, 1974). Codes for teeth as in Figure 6.2. Scale in years after birth.

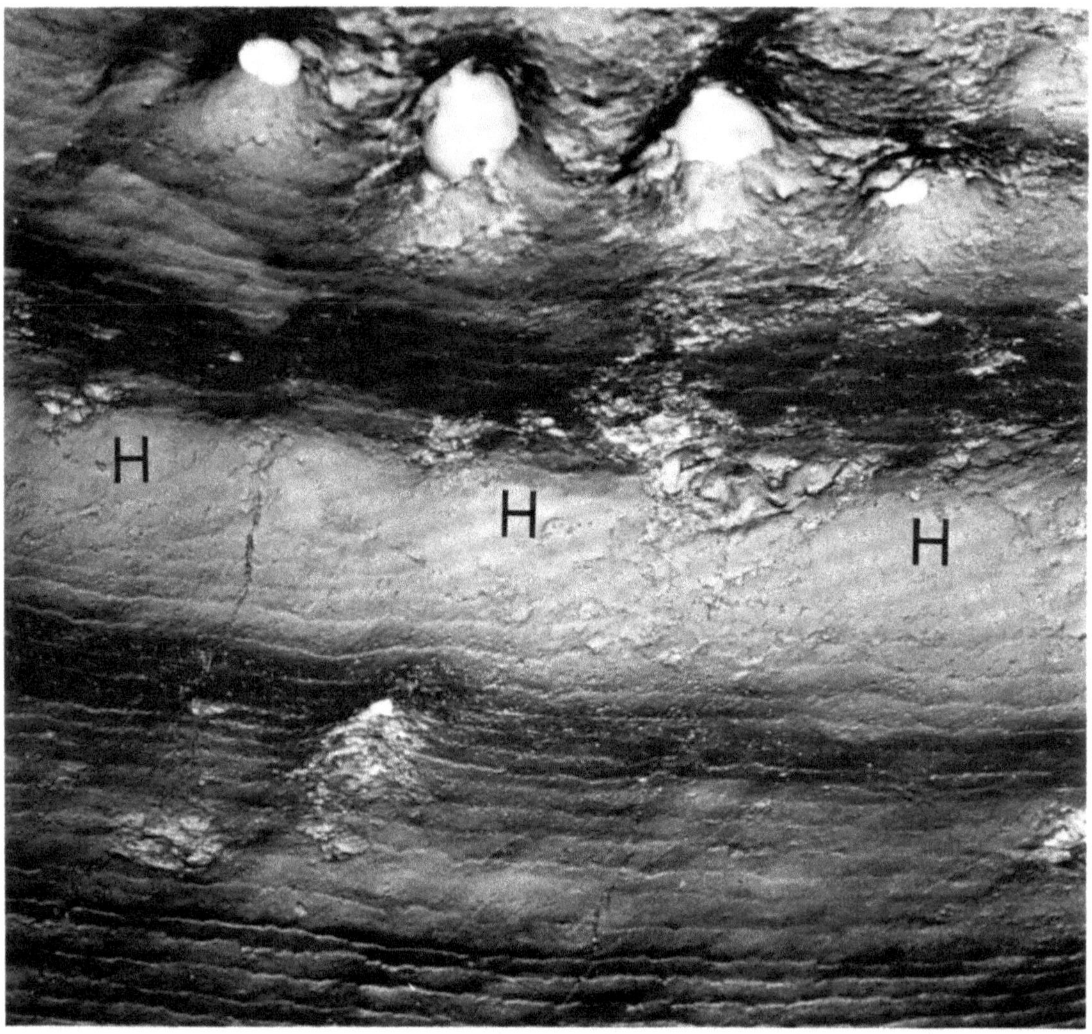

Figure 6.6 Perikymata on the crown surface of a human canine from the Liverpool Street/Broad Street site, London. Material from the Environmental Archaeology Unit, Museum of London. a marked groove-like defect of enamel hypoplasia (marked H-H-H) can be seen, and a pitted defect is parallel to it, towards the top of the field of view. Silicone rubber impression taken from the surface, sputter coated in gold and examined by scanning electron microscopy, using secondary electron mode and inverted video. This is a time-saving technique, which gives the illusion of a positive image to the negative impression, although the shadows are not quite as they would be if the crown surface itself were being examined. Field width c. 1.2μ.

PART 2

Case Studies

7. PATTERNS OF LIFE AND DEATH: THE PALAEODEMOGRAPHY OF MEDIEVAL YORK.

Anne L. Grauer

Introduction

The field of palaeodemography is relatively new. It has developed subsequent to the advances made by skeletal biologists and the growing awareness that human skeletal remains are vital archaeological resources. Palaeodemography has the potential to play a key role in disclosing the intricacies of past life and death patterns. This holds especially true for patterns believed to be manifest in medieval Britain.

Although the field of demography has been used extensively to reconstruct medieval life, most studies have been limited to well-documented periods of medieval history, that is, to the later periods. Many studies, for instance, rely heavily upon parish records (Cowgill, 1963, 1967). But these records were not made compulsory until 1538 (Palliser, 1979:112). Similarly, the use of tombstones did not become a popular means of marking graves with information until the 17th century (Johnson, 1912), thereby also serving as a constraint. While manorial records have been used to recreate the medieval demographic structure of rural villages (Razi, 1980), they are of little assistance in reconstructing the demography of urban centres. Medieval wills and the lay poll tax have also been utilized as tools for understanding medieval life (Bartlett, 1953; Heath, 1984) but "will-making seems to have been a habit acquired by the newly wealthy, merchants and craftsmen in particular, towards the end of the 14th century and once adopted by them became fashionable" (Kermode, 1982:10).

A problem of bias thus appears evident in research concerning medieval life in Britain. The demographic reliance upon written documentation has contributed little toward our understanding of the lives of the urban poor and the lives of women and children. The goal of this paper is to explore the ways in which a palaeodemographic analysis can assist in filling these gaps in our knowledge.

Palaeodemographic Analysis

Palaeodemographic evaluation of skeletal populations has become a primary means of highlighting the presence of stressors. Life table analyses, which include the calculation of mortality rates, survivorship curves and life expectancy rates, led researchers such as Lallo (1973), Green *et al* (1974), Swedlund and Armelagos (1976), Palkovich (1980) and Kennedy (1984) to make significant inferences about patterns of nutritional deficiencies and disease according to age, sex, status, geographical location and subsistence pattern. But life table analysis has its problems. The debate over the applicability and biases inherent in the use of skeletal material has recently become heated (see Angel, 1969; Bocquet-Appel and Masset, 1982, 1985; and Howell, 1982 for arguments against the use of palaeodemography as accurate reflections of archaeological populations, and Van Gerven and Armelagos, 1983; Buikstra and Konigsberg, 1985; and Green *et al*, 1986 for arguments in favour of careful application of palaeodemographic data to archaeological material).

The thrust of the criticism against life table analysis has been aimed at the use of absolute age at death calculations from skeletal material, the inability of the results to reflect mortality patterns of intended populations (the mortality curve mimics the reference population employed to age the skeletal material), the necessity to assume stable population dynamics, and the lack of living populations with similar demographic structure as those created from skeletal populations.

In refutation, these criticisms have been challenged by the use of multiple techniques of ageing and sexing skeletal material (Lovejoy *et al*, 1985), which rely on the concept of developmental ages rather than absolute age at death. Also in refutation of the criticism, Greene *et al* (1986), using Weiss' (1973) model life tables, show that their skeletal samples from Kulubnarti "are more than copies of any so-called reference population. The patterns, instead, are a reflection of one set, among many, of real and biological and cultural processes affecting life and death" (Greene *et al*, 1986:199).

The necessity to assume stable population dynamics has also been addressed by Moore *et al* (1975) and Johansson and Horowitz (1986). Assuming first that a population is stable allows the researcher to explore the effects of possible instability in the population. For instance, a population with high proportions of adult deaths may, assuming that the sample is not biased and that the population is stable, be experiencing low sub-adult mortality. The mortality pattern, however, can be analyzed further by assuming that the pattern might be misleading. Changes in fertility rates and migration may account for the results. Hypothetical increases or decreases in the fertility rate of the population under consideration will concomitantly change the pattern of mortality for the population being treated as a cohort. The mortality pattern may in fact be top-heavy not because of low fertility, high migration of adults into the community or high migration of subadults out of the community. Moore *et al* (1975), Johansson and Horowitz (1986) and Horowitz and Armelagos (1988) suggest that researchers become sensitive to the effect of instability upon a population.

Sensitivity occurs when the researcher asks how the effects of increasing or decreasing fertility rates of changing migration patterns will affect the paleodemographic measures. The results of the initial assessment carried out under the assumption of stable population conditions, therefore, are not adopted as strict indications of the population's dynamics. They are treated as the basis for further exploration.

The final criticism, that results of life table analyses lack correlates to living populations, has been addressed by Greene *et al* (1986). Here the authors show that their results from the Kulubnarti skeletal population appear similar to several living populations.

It appears at this time that palaeodemography can cautiously be used to assess patterns of adaptation and change in archaeological populations. This study will use life tables to examine the presence of stressors in the St. Helen-on-the-Walls population, focussing primarily upon mortality rates, survivorship curves and life-expectancy rates.

The formulation of life tables allows for a critical examination of a population. Life tables consist of seven columns representing the age interval (x), the actual number dead in each age interval (d^1x), the percentage of individuals dying at an age interval out of the total population (dx), the percent living at an age interval (lx), the probability of dying at an age interval (qx), the number of years lived by the total population between age x and x+1 (Lx), and the calculated life expectancy rate at an age interval (e^o_x). Life tables thus use both the mortality rate of the population and the survivorship curve to arrive at a life expectancy rate and mean age of death for the population.

The analyses of mortality rates (dx) are particularly useful to the study of past human populations. Since mortality can be viewed as true maladaptation, age intervals containing high numbers of deaths are especially important to the researcher, for they can elucidate age intervals of great stress to members of the population or can highlight the presence of sample bias. Brothwell (1971), Weiss (1973) and Buikstra (1981) use mortality rate assessments as guides to sample bias. Most common is the potential for infant under-enumeration, which can be detected by dividing the number of infants (0-1 year old) by the number of subadults (1-20 years old) in the population (Brothwell, 1971), or, according to Weiss (1973), which might exist if subadults contribute to less than 30% of the total population.

The survivorship curve (lx) can also serve as a visual clue to palaeodemographic patterns within and between populations. It is constructed by subtracting the proportion dead at each age interval from 100 percent alive at birth. Although survivorship curves do not lend themselves to statistical evaluation, researchers have used them to detect differences in mortality patterns caused by differing conditions during life. As an example, Green *et al* (1974) compare the survivorship curves of individuals buried with cemetery superstructures to individuals buried without superstructures. The patterns, the authors contend, show that burials with superstructures may reflect individuals of higher status.

Life expectancy rates (e^o_x) are also useful inter- and intra-population measures. Several researchers have examined changes in life expectancy rates between populations over time (see Acsadi and Nemeskeri, 1970; Russell, 1948); Brothwell, 1971; Angel, 1969 and Weiss, 1973). These studies have challenged the researchers to explore biological and cultural differences between the populations as possible means of explaining variation. Other researchers (Lallo, 1973; Lallo *et al*, 1978; Green *et al*, 1974) have assessed changes in life-expectancy rates within a population over time. Yet another application has been towards assessing differences in patterns according to differences in status (Green *et al*, 1974; Armelagos *et al*, 1981). These studies suggest that increased status brought increased life-expectancy rates within a population.

The Sample

The medieval cemetery of St. Helen-on-the-Walls (*circa* 1100-1550 A.D.) at York, England, provided three critical requisites for analysis. First, it provided an homogeneous skeletal sample. Documentary sources suggest that the St. Helen-on-the-Walls parishioners were amongst the poorest in the city. Second, all ages and both sexes were present in the sample, suggesting that blatant burial selection criteria were not practiced. Last, the sample size (N = 1014) was large enough to allow statistical evaluation of the trends.

Expectations

Palaeodemography can serve as a means of assessing the presence of stressors within a population without requiring the isolation of the specific stressor at work. Using this premise, several palaeodemographic expectations can be formulated for the St. Helens population. The expectations are contingent upon the assumption that stress within the population will be detected in the mortality pattern. It is also assumed that stressors do not uniformly affect every member of the population, but, rather, depending on the type of stressor, will affect certain age groups more frequently than others. The presence of malnutrition and infection is expected to contribute to a high subadult mortality rate for the St. Helens population due to the particular susceptibility of these age groups to squalid environmental conditions. The stressors, no doubt, will have played an important role in adult life as well, contributing to an expected low mean age at death and low life-expectancy.

Differences in the mortality rates between males and females are also expected. An increased mortality of females during reproductive years is expected due to the compounded physiological disruption caused by malnutrition, infection and pregnancy. Cowgill (1963; 1967) has suggested from 16th century documentary information from the city of York that "although the data indicate no statistically significant difference between the male and female mortality in the York data, there is a difference, possibly due to "chance", but nevertheless occurring as a real feature of the sub-population in question, the female death rate at all times until the end of life being rather higher than the male" (Cowgill, 1967:61).

The palaeodemographic analysis of the St. Helens population was therefore initiated with three goals in mind. The first goal was to utilize palaeodemographic analysis, giving special attention to the criticisms of the techniques. This was achieved, in part, by using multiple ageing and sexing techniques, which were used in order to avoid the potential towards mimicking the age at death pattern of a single reference population and by exploring the applicability of the stable population theory in the analysis and the St. Helen-on-the-Walls population. The second goal was to use palaeodemographic measures as a means of formulating inferences about population processes working in medieval urban centres. The last goal was to assess whether the palaeodemographic patterns were consistent with the expectations.

Methods and Techniques

The palaeodemographic analysis of the St. Helens skeletal population required a determination of sex and age for all individuals in the population. Four techniques were employed to age the skeletal material. The primary technique was dental attriton rates (Miles, 1963a; Brothwell, 1981; Lovejoy, 1985) and dental eruption patterns in juveniles (Schour and Massler, 1941; Hunt and Gleiser, 1955; Miles, 1963b; Moorrees *et al*, 1963a, 1963b; Brothwell, 1981). These techniques were chosen due to the frequently fragmentary nature of the post cranial skeletal material and the comparatively good condition of the dentition. When the post cranial remains were adequately preserved other means of ageing were also used. The degree of epiphyseal union, for instance, was used to assess the age of subadults (McKern and Stewart, 1957; Krogman, 1962; Ubelaker, 1978), while the pubic symphysis was used as an age indicator for adults (Todd, 1920, 1921; McKern and Stewart, 1957; Meindl *et al*, 1985). The last means of assessment used cranial suture closing only when other means of ageing were difficult or impossible.

These four techniques allowed for the eventual seriation of the skeletal material by age. The seriation of the population relied upon the developmental age of the individual and did not assume that archaeological populations could be assigned absolute ages using a modern reference population.

Sex was assessed using three separate techniques, the first requiring a visual evaluation of the pelvis (Brothwell, 1981). The subpubic angle and sciatic notch were used as primary deterinants of sex (Krogman, 1946, 1962; Washburn, 1948; Phenice, 1967, 1969). Sex was also determined using features of the mandible and cranium (Keen, 1950).

Results and Discussion

The excavation of the St. Helen-on-the-Walls cemetery yielded 1014 discretely buried skeletons. Of these, 637 individuals were assigned to age-group intervals, which ranged from one-year intervals up to the age of 5, five-year intervals up to the age of 25, and ten-year intervals up to the maximum age of 65. A sum of 377 skeletons could only be assigned as a "subadult" (0-20 years old) or an "adult" (20-65 years old) due to the severely fragmentary nature of the skeletal material. These individuals were placed in the most refined age categories possible. The following palaeodemographic measures were prorated, using the proportion dying at each age-group interval of those capable of being aged as a reflection of the total population. The clearly aged individuals were used as a guide to the proportionate placement of the more difficult individuals.

a) Population Patterns

Life tables were formulated to assess the mortality pattern and life expectancy rates of the St. Helens population. It was proposed that stressors associated with the St. Helens socio-economic environment would contribute to high infant mortality rates and a generally low life expectancy rate for this population.

The life table for the St. Helens population can be found in Table 7.1. The mortality rate, column dx, which is graphically represented in Figure 7.1, indicates that age-group interval 5-10, in which 11.1% of all deaths occur, age group interval 25-35, in which 23% of all deaths occur and age group interval 35-45, in which 20.1% of the deaths occur, have the greatest mortality rates of any age group intervals. The calculation of life expectancy rates, column e^{o}_{x}, indicates that at birth an individual from St. Helens could expect to live 29.2 more years.

The mortality pattern and the life expectancy rate are perplexing in light of the initial predictions. For instance, the St. Helens population was expected to display high infant mortality rates as a reflection of the susceptibility of children to the effects of malnutrition and infection. However, the mortality rates are not only low for all young subadult age-group intervals up to the age of five in comparison to other age-group intervals, the percentage of subadults versus adults in the St. Helens population is also low. When the St. Helens population is divided into two age groups, subadult (ages 0-20) and adult (ages 20-65), subadults account for only 28% of the total group, a proportion that is low when compared to model life tables (Weiss, 1973). Perplexing, too, is the seemingly high life expectancy rate at birth for the St. Helens population.

Russell (1948) has reconstructed life expectancy patterns for medieval England using documentary evidence, thus recreating the demographic structure of the wealthy. With a maximum age at death of 90 years, Russell finds that the expectation of life at birth ranges from a high of 35.3 years before 1276, to a low of 17.3 years during the plague years of 1346-1375. This averages to a life expectancy of 26.3 for years between 1276 and 1450. Weiss (1973) suggests that the "deficit in juveniles acknowledged by Russell would tend to raise the mortality rates and lower the life expectancies still further" (Weiss, 1973:42), increasing the disparity between the St. Helen life expectancy at birth and those calculated by Russell.

It is when the St. Helens life table data are compared to model life tables that the discrepancies and patterns can be assessed. Comparisons between the St. Helens life table and Weiss' (1973) model life tables highlight three inconsistencies within the St. Helens life table data. Weiss asserts that 49 of the 72 model life tables follow a predictable mortality pattern. First, they indicate that the probability of dying (column "qx") within the first year of life is greater than the probability of dying between the ages of 15-20. Second, the probability of dying between the ages of 15-20 is greater than dying between the ages of 10-15. And last, that the proportion of juvenile deaths in the population is greater than 30% of all deaths. Calculated life tables unable to satisfy these conditions should be "considered as unlikely to model real stationary conditions and should be used only if there is compelling evidence to do so" (Weiss, 1973:47).

The results of the St. Helens life table fail to satisfy any of these criteria. As column "qx" of Table 7.1 indicates, the probability of dying at the age group interval 0-1 is .026, which is considerably lower thanthe probability of dying at the age group interval 15-20 (qx = .078). Similarly, the probability of dying at age 10-15 (qx = .081) is higher than the probability of dying between the ages of 15-20. The proportion of subadult deaths in the St. Helens population (28%) also suggests that the population may not be stable.

Since it appears from the comparison of the St. Helens life table to Weiss' model life tables that St. Helens cannot be treated as a stable population, it becomes obvious that treating the St. Helens population's mortality pattern and life expectancy rate at birth as realistic is unjustified. However, it does not imply the resulting patterns are useless towards the understanding of the demographic processes occurring in this medieval population.

Several causes for the proportionately low numbers of subadults in the St. Helens population were tested. The first proposed cause of bias was the presence of differential skeletal preservation according to age. This proposition was based upon the recognition that subadult skeletal material is more fragile than adult material and less likely to be preserved. An archaeologist therefore has a greater chance of exhuming an older-aged individual than a younger-aged individual. However, the results of the St. Helens excavation do not support this contention. The age-group interval 5-10, for instance, displays a greater number of individuals than the age-group interval preceding it. This result suggests that preservation is not merely a function of age or bone size.

The second proposed cause for subadult skeletal bias considered differential burial practices as a function of age. This hypothesis was based upon the Christian practice of excluding unbaptized infants from graveyard burial or isolating these burials on the north side of the graveyard or church structure (Johnson, 1912). Myrc (1868), however, suggests that medieval churchmen were concerned about infants dying before baptism could be performed. Subsequently, midwives and other laymen were taught how to baptize newborn infants believed too weak to survive. Berry and Schofield (1971), in their study of age at baptism in pre-industrial England, have concluded that baptisms between the period of 1650-1700 took place within the first month of life. Therefore, in the analysis of burials in the St. Helens population infants aged 0-1 would be used to test whether baptism played a role in mortuary behaviour.

If the St. Helen-on-the-Walls community chose to include but segregate unbaptized infants, one area would display a disproportionately higher number of infant burials. To test for this occurrence the north side of the cemetery was segregated from the other areas of the cemetery. A Chi-square test (Table 7.2) was performed to assess whether the north side had a disproportionate number of infants aged 0-1 years old compared to subadults 1-20 years old. The test indicated that the differences were not statistically significant at the .05 level.

On the other hand, if unbaptized children were excluded from graveyard burial then the demographic structure of the cemetery population would appear to have a disproportionately low number of individuals in the first age-group interval. The mortality rate for the St. Helens population indicates that 2.6% of the population died before the age of one year, contributing to a low probability of dying (qx = .026). While the probability of dying is the highest of all age-group intervals below the interval of 5-10 years old, it is very low when compared to Acsadi and Nemeskeri's (1970) and Weiss' (1973) model life tables, which indicate the probability varying between .1 to .4. This may suggest that the low percentage of subadults represented in the age-group interval 0-1 in the St. Helens population was caused by burial practices, which excluded the burial of unbaptized children within the confines of the church cemetery.

The results of the tests for subadult under-enumeration are therefore revealing. They suggest that the bias against subadults in the St. Helens population was not necessarily due to archaeological preservation biases. Rather, it appears that other processes are contributing to the low percentage of subadults, one of which may be the exclusion of unbaptized infants from the cemetery sample.

However, the role of adults in the St. Helens population must not be overlooked at this juncture. While archaeologists and physical anthropologists must be sensitive to the effects of preservation biases against subadult skeletal material, the potential for adult over-enumeration in the skeletal population must also be considered. Medieval historians such as Russell (1948) propose that urban communities were maintained by high rates of adult immigration. If this proposition is true, then the skeletal population from St. Helens may be displaying an over-enumeration of adults, rather than simply an under-enumeration of subadults.

Assessing the numbers of individuals migrating into York is difficult. So, too, is assessing the proportion of immigrants into the St. Helen-on-the-Walls community. According to Russell (1948), the Freemen's Rolls for the city of York suggest that 25% of the recorded individuals were originally from towns located five to fifty miles outside of York. This large proportion, if assumed to have migrated to York as adults, would skew the results of the life expectancy rate at birth calculations for the St. Helens population, rendering the mean age at death of 29.2 years old far too high.

The results may be further exacerbated by unequal migration patterns within York's parishes. It must be remembered that York's parishes differed socio-economically from one another. The St. Helen-on-the-Walls parish, according to the Lay Poll Tax Returns, appears to have been on the low end of the scale. The parishioners were capable of paying only the smallest amount of tithes when compared to the payments paid by members of other parishes in the city's centre. Therefore, when using documentary evidence to assess the amount of immigration into the city and the assessed wealth of a parish, the researcher must recognize that the records were often biased in favour of York's more prosperous citizens. Migrants, coming to the city in search of work, may have been frequently omitted from the Poll Tax records and may have constituted a greater percentage of the poorer parishes than of the wealthier parishes, which are well represented in documentary evidence.

Further examination of the migration patterns into York is essential towards the understanding of the city's population processes. The migration of adults into a community not only skews the percentages between juveniles and adults in the population, but may have a profound effect on fertility patterns. Both Russell (1948) and Goldberg (1986) have suggested that a need for personal economic security delayed the age of marriage in urban centres. This delay may have influenced fertility rates within the city. If the age of marriage was delayed, the number of offspring may have been reduced. For the St. Helens population this would have contributed to a low growth rate, or a negative growth rate if the birth rate of the population could not balance the mortality rate.

The presence of low or negative fertility rates also contradicts the life table assumption of population stationarity. Taking heed of the discussions of Moore *et al* (1975), Johansson and Horowitz (1986) and Horowitz and Armelagos (1988), concerning the effects of changing growth rates on life table results, the response of the St. Helens population life table to low or negative fertility rates was explored. According to these authors, in a stable population the mean age at death will decrease as the fertility rate of the population increases. When the St. Helens population is hypothetically treated as stable (migration into the community equals the migration rate out of the community or no migration is occurring in either direction) and the relationship between the St. Helens mean age at death and life expectancy rate is altered to assume a +1/2% rise in the growth rate, it appears that while the life expectancy rate remains constant, the mean age at death will decrease from 29.2 years to approximately 25 years. Conversely, if the St. Helens population underwent periods when replacement of the population through births was impossible, as suggested by historians, then the mean age at death would increase to approximately 36 years old, assuming a negative growth rate of -1%.

It may therefore be possibel that three processes were taking place simultaneously in the St. Helens population, causing what appears to be an under-enumeration of subadults in the community and an artifically high life expectancy rate at birth. The first process may have been the exclusion of unbaptized infants from the church graveyard. The second may have been a high rate of adult immigration into the community, skewing the cemetery sample towards older individuals. The last process contributing to the life table results may have been a low fertility rate within the city, which would result in an increased mean age at death for the population.

There seems little doubt that the city of York suffered from many fluctuations in population size and proportion. Changes in York's economy, coupled with recurring bouts of pestilence and plague, caused the population of St. Helen-on-the-Walls parish to vary throughout the medieval centuries. Periods of population increase were punctuated with events that caused immediate decreases in size. However, a more general picture of York's population between the 12th to the mid 16th centuries shows an increase in size. I believe that this overall increase was due, in part, to a high rate of adult migration into the city, which, by creating population instability, masks the potential for more specific evaluations of fertility rates. Obviously, I propose that these same processes were occurring within the St. Helen-on-the-Walls parish and contributed to the unusually low subadult mortality rate and probability of dying, and thus the particularly high life expectancy rate at birth in the St. Helen-on-the-Walls population.

b) Gender Patterns

Different demographic patterns based upon gender were also expected to be found in the St. Helens skeletal population. This proposition stemmed from Cowgill's (1963; 1967) suggestion that 16th century parish documents from the city of York showed a higher female mortality rate than male mortality. The results of the St. Helens investigation indicate that mortality rates by gender differ (Table 7.3 and Figure 7.2). The female mortality curve appears to rise to a peak at age-group interval 25-35 and steadily decreases at older age-group intervals. The male mortality curve rises to a peak at age-group interval 35-45 and then decreases steadily. The life expectancy rate upon reaching adulthood (age 20) also differs by gender. Females reaching the age-group interval 20-25 can be expected to live 18.3 more years, while males reaching the same age-group interval can expect to live 20.6 more years. The Kolmogorov-Smirnov test indicates that the differences between these two patterns are statistically significant (Table 7.4).

When these results are compared to model life tables the patterns displayed in the St. Helens population become even more intriguing. According to the model life tables constructed by Coale and Demeny (1983) females consistently display a higher life expectancy rate than males. In the St. Helens population males clearly have a greater life expectancy rate than females, which is graphically evident in their different survivorship curves (Figure 7.3). Here it can be seen that at no time does female survivorship surpass that of males.

Two explanations for the gender patterns of mortality are plausible. The first cause may be the presence of a high number of maternal deaths. However, female mortality appears to peak at the age interval of 25-35, a late age-group interval if historians are correct in their premise that females married in their teens (Shahar, 1983). The second cause may be due to migration patterns. I suggest that both of these factors play a role in the St. Helens gender mortality patterns.

In a recent article by Goldberg (1986), the traditional view of female marriage and migration patterns, which maintains that medieval women married before the age of twenty and were unlikely to leave the home of their parents unless married, is challenged. She asserts that medieval cities and towns frequently show a higher proportion of females than males within the population. "Using poll tax returns of 1377 as a source only 90.5 adult males excluding servants are found to every 100 women in York, only 86.1 males in Hull and as few as 85.4 in Carlisle" (Goldberg, 1986:19). These low sex ratios are the result of large numbers of females migrating to the city as a response to the shift from arable to pastoral land use and the depletion of labour in the cities after the plague. Migration, therefore, of females as well as males, played a role in marriage and fertility patterns:

"The argument here is that marriage in the later fourteenth and fifteenth centuries was relatively late, that is, frequently in the mid-twenties and very rarely in the 'teens even for women and also companionate in the sense that the age difference between spouses was not very great. For women the opportunity for employment, in service or as independent traders, created by the years of economic expansion and labour shortage following the Black Death allowed them to delay marriage... Above all, it permitted them a degree of independence that younger girls living at home with their parents could not have enjoyed (Goldberg, 1986:25-26).

This pattern may have affected the St. Helens population. When the proportion of females to males is examined, females constitute 53.6% of the adult population. This trend cannot be attributed to biases of skeletal sexing techniques, as the incorporation of several sexing techniques circumvented this tendency noted by Weiss (1973) and Meindl *et al* (1985). Although these differences were not found to be statistically significant using a Chi-Square test (Table 7.5), this proportion is perplexing and may indicate that a disproportionate number of women were migrating into the community.

The proposed pattern of female migration and late age of marriage also serves to explain the peak in female mortality at the age interval 25-35. A delay in marriage age can be associated with a delayed age of maternal stress. The age-group interval 25-35 does not appear to be too late to be considered indicative of the presence of maternal stress if the average age of marriage in the population falls nearer to the age of 20 than to the age of 17, and the traditional view that childbirth infrequently occurred outside marriage is adopted.

However, a greater proportion of females migrating into the city of York and a delayed age of marriage (and thus maternity) does not in itself explain the unusual pattern of differing life expectancy rates between the genders. Galloway (1985) has suggested from 17th and 18th-century London data that mid-adulthood peaks reflect migration patterns into cities and the exposure of new individuals to the cities' pathogenic environment. Perhaps a greater proportion of the females in the St. Helens population were migrants, who, unlike the greater proportion of indigenous males, were exposed to new stressors alongside the stressors associated with childbirth. Thus when migration patterns are coupled with marriage patterns for the city of York, the difference between female and male mortality patterns may be explained by adding maternal stresses to the other stresses encountered by women reaching the age of 25.

Conclusion

This chapter had three goals. The first goal was to circumvent some of the criticism aimed at discrediting the use of palaeodemography. The use of multiple ageing techniques and seriation rendered the replication of a single reference population unlikely. The seriation of the skeletal material provided an assessment of the developmental age of individuals in the population rather than the absolute age, thus minimizing the tendency towards "mimicking" reference populations as suggested by Bocquet-Appel and Masset (1982; 1985).

The second goal was to use palaeodemographic measures to make inferences about past mortality and morbidity patterns. Life tables were constructed for this purpose. The results indicated that the St. Helens population displayed a low mortality rate and a low probability of dying for individuals between birth and five years old, with an abrupt increase at the age-group interval 5-10. The mortality rate and probability of dying peaked once again between the age-group intervals 25-35 and 35-45. The life expectancy rate at birth appeared to be 29.2 years.

When the results of the St. Helens life tables were compared to model life tables constructed by Weiss (1973), the results were re-evaluated. It was suggested that three processes were responsible for the life table results: 1) that unbaptized infants were excluded from the cemetery; 2) that the underenumeration of subadults was exacerbated by an overenumeration of adults who had migrated into the community; and 3) that a delayed age at marriage could have dampened the fertility rate, thereby causing the mean age at death for the St. Helens population to appear high.

Gender patterns were also explored. The results indicated that the peak in female mortality occurred at the interval of 25-35, while male mortality peaked at 35-45. These differences proved statistically significant. Also intriguing was the increased male life expectancy rate (20.6 years), while females reaching adulthood could only expect to live 18.4 more years. The reasons for this trend could not be explained by assuming that the St. Helens population was not stable. Rather, it was suggested that a greater proportion of females in the St. Helens population were migrants compared to the males, and that the abrupt introduction of new stressors associated with medieval urban environments and the added stressors associated with childbirth decreased the likelihood of females living as long as males.

The third goal of this chapter was to examine whether the St. Helen results complied with the initial expectations for the group. It was proposed that the St. Helens population, by virtue of its low socio-economic position within the city of York, would suffer from malnutrition and infection. The presence of these stressors would be inferred from high subadult mortality rates and a generally low life expectancy rate for the population. Differences between male and female mortality rates and life expectancy rates were also expected. Women were expected to show a greater mortality at younger ages and a lower life expectancy rate than males.

The results of the palaeodemographic analysis did not comply with the initial expectations. At first glance subadult mortality appeared low and the life expectancy rate for the population was high when compared to other medieval populations. However, the roots of the discrepancies were not associated with an inadequate assessment of the St. Helens population socio-economic condition or with an inadequate assessment of the effects of low status upon demographic processes, but, rather, could be explained by treating the population as non-stable. Hence, while many expectations were not met by the data, the paleodemographic evaluation of the St. Helen-on-the Walls skeletal material provided insight into the biases and processes at work within the city of York.

References

Acsadi, G.Y. and Nemeskeri, J. (1970). *History of Human Life Span and Mortality.* Budapest: Hungarian Academic Society.

Angel, L. (1969). The bases of paleodemography. *Am. J. Phys. Anthropol. 30*:427-38.

Armelagos, G.J., Goodman, A.H. & Bickerton, S. (1980). Determining nutritional and infectious disease stress in prehistoric populations (abstract). *Am. J. Phys. Anthropol. 52*:210.

Armelagos, G.J., Jacobs, K.H. & Martin, D.L. (1981). Death and demography in prehistoric Sudanese Nubia. In *Mortality and Immortality: The Anthropology and Archaeology of Death.* eds. S.C. Humphreys and H. King. London: Academic Press.

Bartlett, N. (1953). The Lay Poll Tax Returns for the city of York in 1381. *East Riding Hist. Trans.* (offprint).

Berry, B.M. and Schofield, R.S. (1971). Age at baptism in pre-industrial England. *Pop. Studies 25*:453-63.

Bocquet-Appel, J. and Massett, C. (1982). Farewell to paleodemography. *J. Hum. Evol. 25*:321-33.

Bocquet-Appel, J. and Massett, C. (1985). Matters of moment. *J. Hum. Evol. 14*:107-111.

Brothwell, D. (1971). Palaeodemography. In *Biological Aspects of Demography.* ed. W. Brass. pp. 111-30. London: Taylor and Francis.

Brothwell, D. (1981). *Digging Up Bones.* 3rd edition. London: British Museum (Natural History).

Buikstra, J. (1981). Mortuary practices, paleodemography and paleopathology: a case study from the Koster site (Illinois). In *The Archaeology of Death* eds. R. Chapman, I. Kinnes, & K. Randsborg. pp. 123-32. Cambridge: Cambridge University Press.

Buikstra, J. and Konigsberg, L. (1985). Paleodemography: critiques and controversies. *Am. Anthropol. 87*:316-33.

Coale, A.J. and Demeny, P. (1983). *Regional Model Life Tables and Stable Populations* 2nd edition. Orlando: Academic Press.

Cowgill, U. (1967). Life and death in the sixteenth century in the city of York. *Pop. Studies 21*:53-62.

Cowgill, U. and Hutchinson, G.E. (1963). Differential mortality among the sexes in childhood and its possible significance in human evolution. *Proc. Natl. Acad. Sci. 49*:425-29.

Galloway, P.R. (1985). Annual variation in death by age, death by cause, prices, and weather in London 1670-1830. *Pop. Studies 39*:487-505.

Goldberg, P.J.P. (1986). Female labour, service and marriage in the late mediaeval urban north. *North. Hist. 12*:18-38.

Green, S, Green, S. & Armelagos, G.J. (1974). Settlement and mortality of the Christian Site (1050 A.D.-1300 A.D.) of Meinarti (Sudan). *J. Hum. Evol. 3*:297-316.

Greene, D.L., Van Gerven, D.P. & Armelagos, G.J. (1986). Life and death in ancient populations: bones of contention in paleodemography. *Hum. Evol. 1*:193-207.

Heath, P. (1984). Urban piety in the later middle ages: the evidence of Hull wills. In *The Church, Politics and Patronage in the Fifteenth Century.* ed. B. Dobson. pp.Gloucester: Alan Sutton Press.

Horowitz, S. and Armelagos, G.J. (1988). On generating birth rates from human skeletal remains. *Am. J. Phys. Anthropol. 76*:183-188.

Howell, N. (1982). Village composition implied by a paleodemographic life table: the Libben site. *Am. J. Phys. Anthropol. 59*:263-70.

Hunt, E.E. and Gleiser, I. (1955). The estimation of age and sex of preadolescent children from bones and teeth. *Am. J. Phys. Anthropol. 13*:479-88.

Johansson, S. and Horowitz, S. (1986). Estimating mortality in skeletal populations: influences of growth rate on the interpretation of levels and trends during the transition to agriculture. *Am. J. Phys. Anthropol. 71*:233-50.

Johnson, W. (1912). *Byways in British Archaeology.* Cambridge: Cambridge University Press.

Keen, J.A. (1950). A study of the differences between male and female skulls. *Am. J. Phys. Anthropol. 8*:65-79.

Kennedy, K.A. (1984). Growth, nutrition and pathology in changing paleodemographic settings in South Asia. In *Paleopathology at the Origins of Agriculture.* eds. M. Cohen and G.J. Armelagos. pp. Orlando: Academic Press.

Kermode, J. (1982). The merchants of three northern English towns. In *Profession, Vocation and Culture in Later Medieval England.* ed. C. Clough. pp. Liverpool: Liverpool University Press.

Krogman, W.M. (1946). The skeleton in forensic medicine. *Proc. Inst. Med. 16*:154.

Krogman, W.M. (1962). *The Human Skeleton in Forensic Medicine.* Springfield, Illinois: C.C. Thomas.

Lallo, J.W. (1973). *The Skeletal Biology of Three Prehistoric American Indian Societies from Dickson Mounds.* Ph.D. dissertation, University of Massachusetts, Amherst.

Lallo, J.W., Armelagos, G.J. & Rose, J.C. (1978). Paleoepidemiology of infectious disease in the Dickson Mounds population. *Med. Coll. Virginia Q. 14*:17-23.

Lovejoy, C.O. (1985). Dental wear in the Libben population: its functional pattern and role in the age determination of adult skeletal age at death. *Am. J. Phys. Anthropol. 68*:47-56.

McKern, T.W. and Stewart, T.D. (1957). Skeletal age changes in young American males. *Technical Report EP-45.* Natick, MA: Quartermaster Research and Development Center, Environmental Protection Research Division.

Meindl, R., Lovejoy, C.O., Mensforth, R. & Carlos, L.D. (1985). Accuracy and direction of error in the sexing of the skeleton: implications of paleodemography. *Am. J. Phys. Anthropol. 68*:79-85.

Miles, A.E.W. (1963a). The dentition in the assessment of individual age in skeletal material. In *Dental Anthropology.* ed. D. Brothwell. pp.191-209. Oxford: Pergamon Press.

Miles, A.E.W. (1963b). Dentition in the estimation of age. *J. Dent. Res. Supplement (1) 42*:255-63.

Moore, J.A., Swedlund, A. & Armelagos, G.J. (1975). The use of life tables in paleodemography. *Am. Antiq. Mem. 30*:57-70.

Moorees, C., Fanning, E. & Hunt, E. (1963a). Age variation and formation stages for ten permanent teeth. *J. Dent. Res. 42*:1490-1502.

Moorees, C., Fanning, E. & Hunt, E. (1963b). Formation and resorption of three deciduous teeth in children. *Am. J. Phys. Anthropol. 21*:205-213.

Myrc, J. (1868). In *Instructions for Parish Priests*. ed. E. Peacock, from Cotton Ms. and Claudius A II. London: Trubner and Co.

Palkovich, A.M. (1980). *Pueblo Population and Society: Arroyo Hondo Skeletal and Mortuary Remains.* Arroyo Hondo Archaeological Series, Vol. 3.

Palliser, D.M. (1979). *Tudor York.* Oxford: Oxford University Press.

Phenice, T.W. (1967). A newly developed visual method of sexing the *os pubis*. *Am. J. Phys. Anthropol. 30*:297-302.

Phenice, T.W. (1969). An analysis of the human skeletal material from burial mounds in North Central Kansas. *University of Kansas Publication in Anthropology, No. 1.* Lawrence, KS.

Razi, Z. (1980). *Life, Marriage and Death in a Medieval Parish.* Cambridge: Cambridge University Press.

Russell, J.C. (1948). *British Medieval Population.* Albuquerque: University of New Mexico Press.

Schour, I. and Massler, M. (1941). The development of the human dentition. *J. Am. Dent. Assoc. 28*:1153-60.

Shahar, S. (1983). *The Fourth Estate: A History of Women in the Middle Ages.* Trans. C. Galai. New York: Methuen.

Swedlund, A. and Armelagos, G.J. (1976). *Demographic Anthropology.* Dubuque, Iowa: Wm. C. Brown.

Todd, T.W. (1920). Age changes in the pubic bone: I. the male white pubis. *Am. J. Phys. Anthropol. 3*:285-334.

Todd, T.W. (1921). Age changes in the pubic bone. *Am. J. Phys. Anthropol. 4*:1-70.

Ubelaker, D.H. (1978). *Human Skeletal Remains.* Chicago: Aldine Press.

Van Gerven, D.P. and Armelagos, G.J. (1983). "Farewell to paleodemography"? Rumors of its death have been greatly exaggerated. *J. Hum. Evol. 12*:353-60.

Washburn, S.L. (1948). Sex differences in the pubic bone. *Am. J. Phys. Anthropol. 6*:199-208.

Weiss, K.M. (1973). Demographic models for anthropology. *Am. Antiq. 38*, Part 2, Memoir 27.

TABLE 7.1

LIFE TABLE FOR THE TOTAL POPULATION

Age	d`x	dx	lx	qx	Lx	Tx	$e^o x$
0-1	26	2.6	100	.026	98.7	2920.5	29.2
1-2	14	1.4	97.4	.014	96.7	2821.8	29.0
2-3	19	1.8	96	.018	95.1	2725.1	28.4
3-4	21	2.1	94.2	.022	93.2	2630	28.0
4-5	19	1.9	92.1	.021	91.2	2536.8	27.5
5-10	113	11.1	20.2	.123	423.3	2445.6	27.1
10-15	65	6.4	79.1	.081	379.5	2022.3	25.6
15-20	58	5.7	72.7	.078	349.5	1642.8	22.6
20-25	42	4.1	67.0	.061	324.8	1293.3	19.3
25-35	233	23.0	62.9	.366	514	968.5	15.4
35-45	204	20.1	39.9	.504	298.5	454.5	11.4
45-55	143	14.1	19.8	.712	127.5	156	7.9
55-65	57	5.7	5.7	1.0	28.5	28.5	5
	1014						

TABLE 7.2

ASSOCIATION BETWEEN BURIAL AREA AND AGE AT DEATH FOR SUBADULTS

	North	All Other Areas	TOTAL
0-1	12	15	27
1-20	119	189	308
	131	204	

$/x^2/ = .381 < x^2.05 = 3.84$

TABLE 7.3

LIFE TABLES BY GENDER

FEMALES

Age	d`x	dx	1x	qx	Lx	Tx	$e^o x$
20-25	19	.067	100	.067	483.25	1838.75	18.4
25-35	112	.392	93.3	.420	737	1355.5	14.5
35-45	74	.26	54.1	.48	411	618.5	11.4
45-55	61	.214	28.1	.761	174	207.5	7.3
55-65	19	.067	6.7	1.0	33.5	33.5	5
	285						

MALES

Age	d`x	dx	1x	qx	Lx	Tx	$e^o x$
20-25	12	.05	100	.05	487.5	2062.5	20.6
25-35	67	.27	95	.28	815	1575	16.5
35-45	89	.36	68	.53	500	760	11.2
45-55	55	.22	32	.68	210	260	8.1
55-65	24	.1	10	1.0	50	50	5
	247						

TABLE 7.4

KOLMOGOROV-SMIRNOV TEST FOR DIFFERENCES BETWEEN AGE AT DEATH RATES BY GENDER

MALES			FEMALES		
Age	f	cum.	f	cum.	D
20-25	12	.049	19	.067	.018
25-35	67	.32	112	.46	.14**
35-45	89	.68	74	.72	.04
45-55	55	.9	61	.934	.034
55-65	24	1.0	19	1.0	0.0

$D_{.05} = .118$
** Statistically significant

TABLE 7.5

PERCENTAGE OF ADULT MALES AND FEMALES IN THE TOTAL POPULATION

	f	%
Females	285	53%
Males	247	47%
	532	

$/x^2/ = 2.70 < x^2.05. = 3.84$

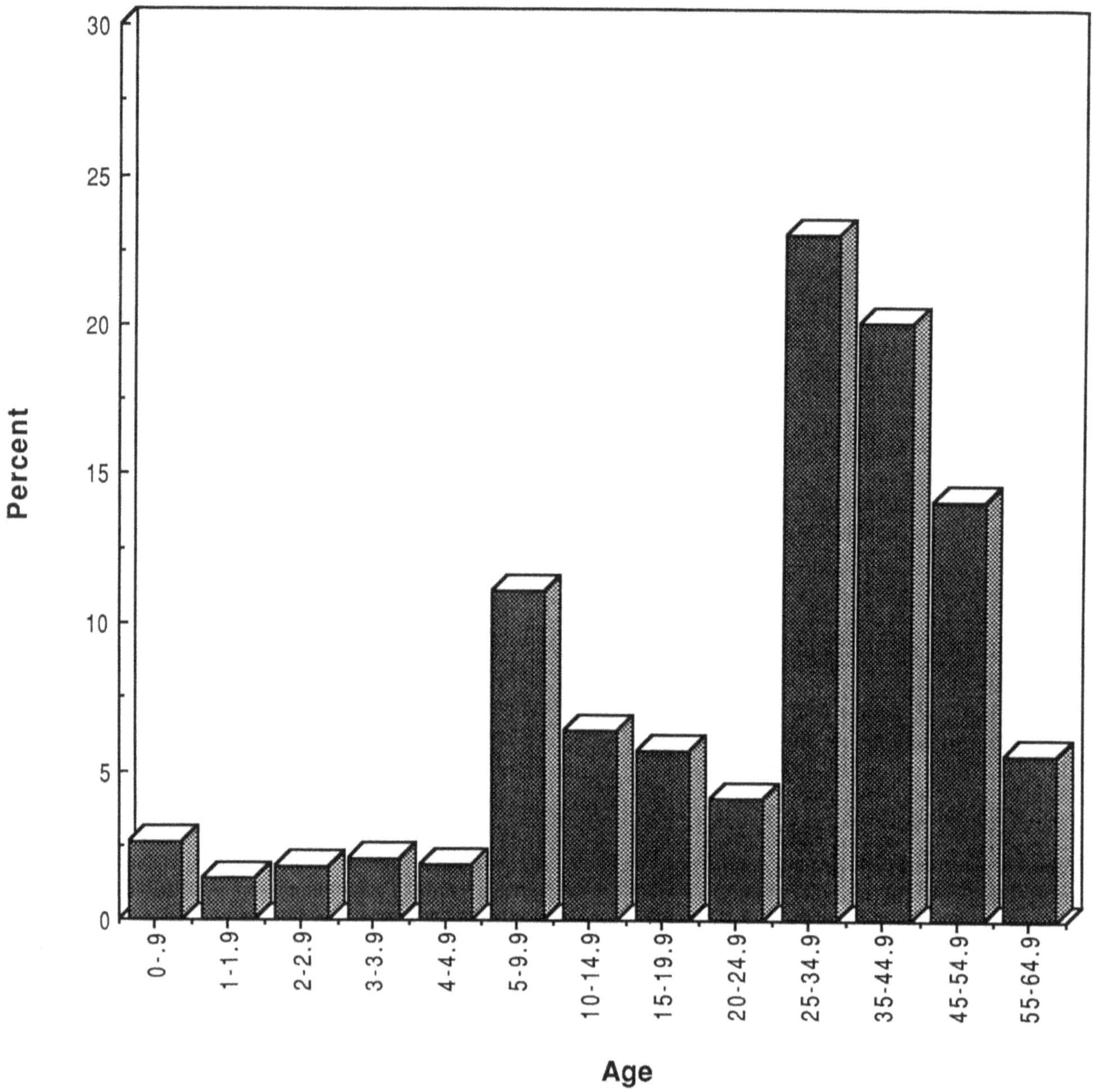

Figure 7.1 Mortality rate for the St. Helen-on-the-Walls population.

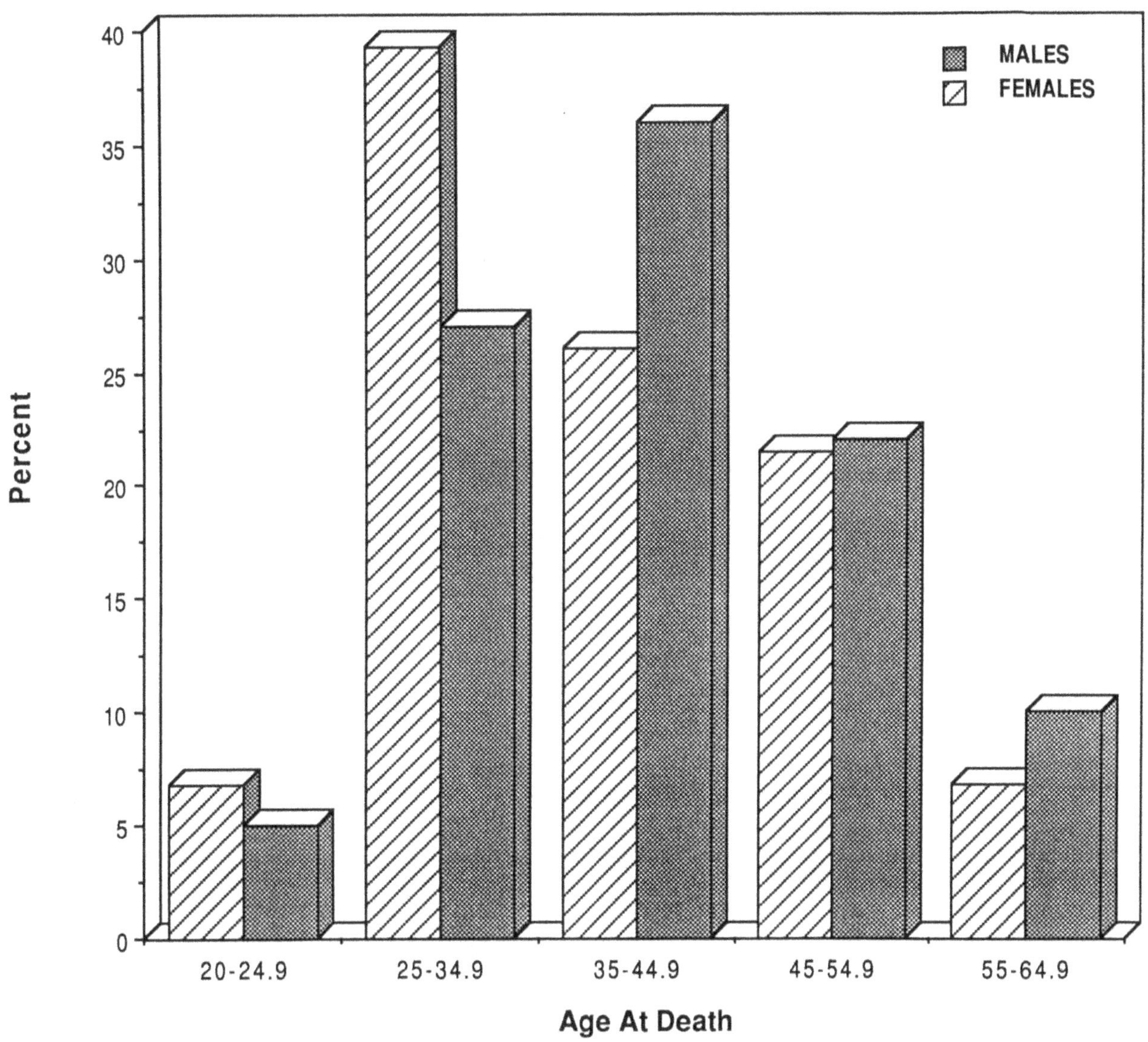

Figure 7.2 Mortality rate by sex in the St. Helen-on-the-Walls population.

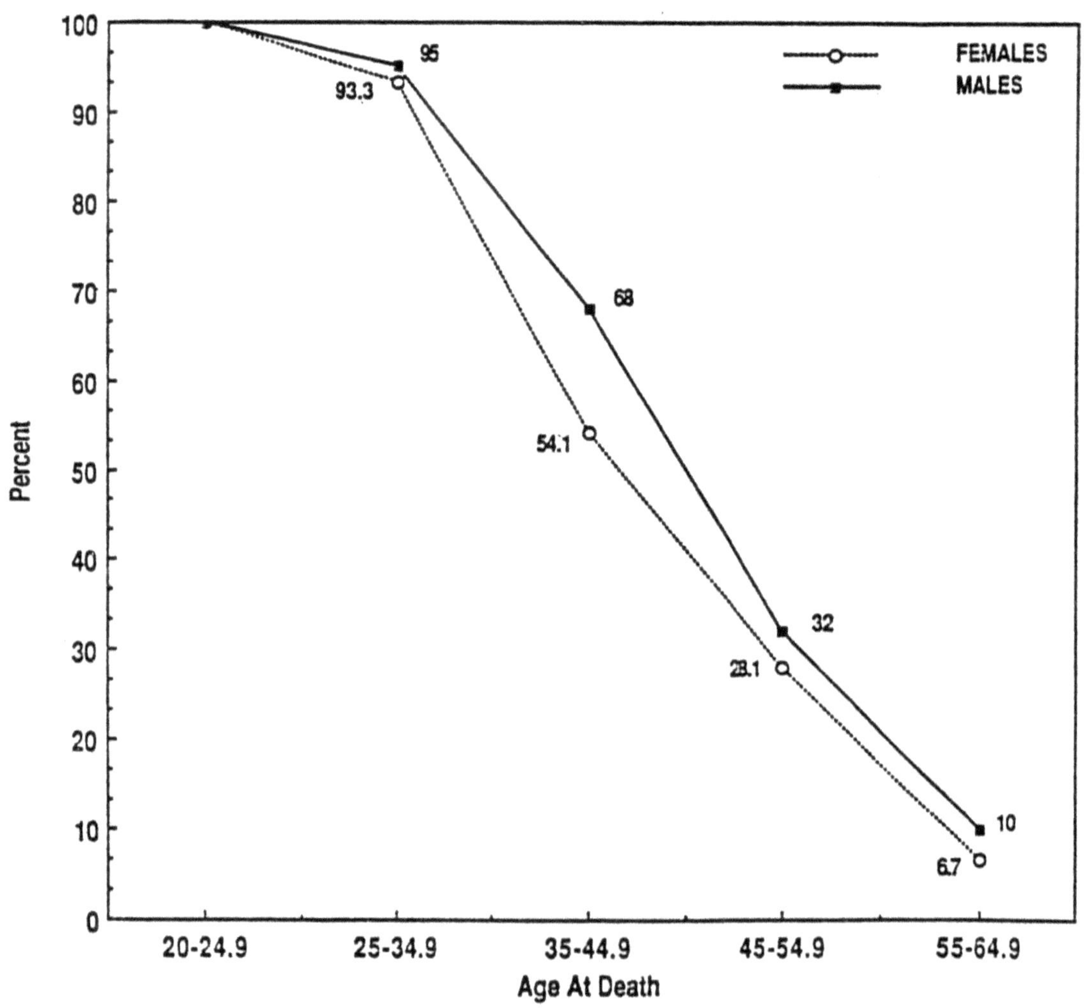

Figure 7.3 Survivorship curve by sex in the St. Helen-on-the-Walls population.

8. EPIDEMIOLOGICAL STUDIES OF DENTAL ENAMEL HYPOPLASIAS IN MEXICO AND BRADFORD: THEIR RELEVANCE TO ARCHAEOLOGICAL SKELETAL STUDIES.

Keith Dobney and Alan Goodman.

Introduction

A living organism is constantly involved in a dynamic interaction with its immediate environment. Imbalances which occur in this equilibrium may result in the occurrence of a variety of pathological conditions; dietary imbalances, for example, can result in a variety of conditions associated with malnutrition.

Defects of the teeth, and associated dental tissues, have long been utilized in both anthropological and archaeological studies as a means of directly or indirectly attempting to resolve questions of diet and general health status (Dobney, 1983). The mouth forms a distinct zone which is in direct contact with the environment. It acts as an entrance for food, is the area of mastication and provides access for parasites and other harmful toxins. The teeth specifically come into contact with every part of the diet and as a result are often physically scarred by this contact. They can be affected by a variety of factors during formation which may leave a permanent record on the enamel or crown surface.

Of the varied quantitative lesions of the enamel which are commonly found, those which form as a direct result of developmental problems have been utilized the most as a retrospective means of observing and recording patterns of past developmental stress. The term enamel hypoplasia simply relates to a deficiency in enamel thickness due to ameloblast disfunction occurring during enamel matrix formation (Sarnat and Schour, 1941). Commonly, the hypoplastic defect is manifested as single or multiple pits, as linear grooves running horizontally across the enamel surface and even as whole areas of missing enamel (FDI, 1982). These quantitative defects contrast with the other major enamel defect known as enamel opacities (or hypocalcifications) which can be formed during both enamel matrix formation and maturation (Goodman and Rose, in press).

A "memory" of past events which has affected ameloblast activity is possible since enamel (unlike most biological tissue) has no capacity to remodel, so once formed, lesions are locked into the enamel subject only to the dictates of physical or chemical erosion. Also enamel, in common with other dental tissues, is formed in very rhythmic biological episodes which have been studied widely (Beynon and Wood, 1987; Bromage and Dean, 1985). The ring-like structures which result also provide a detailed chronology of the "stressors" which affect ameloblast activity (Goodman and Armelagos, 1988).

The presence and frequency of dental enamel defects (DDEs) have been used by human osteologists as an index of the health status of archaeological populations for some years. However, many assumptions and generalisations have been made in the archaeological literature regarding the aetiological processes which result in these defects. These conclusions cannot be substantiated or resolved by the study of archaeological assemblages alone. Direct comparisons of statistical data generated from such studies have been undertaken within and between skeletal assemblages, often without due reference to the modern dental researchers' continuing plight in understanding the myriad aetiological factors that may be influencing the occurrence of DDEs.

There is by no means a clear concise picture to be gleaned from the quite prolific modern epidemiological research carried out over the past 30 years into causal factors involved in the production of DDEs. Table 8.1 includes the main results of a number of these studies undertaken in both developed and underdeveloped communities of the world. As can be seen, there is quite a notable diversity in the frequency of defects. However, care must be taken when comparing data between studies. One reason is that some of the data refers to the deciduous dentition. Another major problem is the differences in method of classification of the various types of DDEs between workers. As Jontell and Linde (1986) point out, the clinical registration of enamel hypoplasias can frequently be difficult to perform and has been based on different diagnostic methods; the differences between different clinical investigators are also very important. These points are equally as pertinent to the study of archaeological assemblages as they are to recent clinical work.

Recording the frequency, severity and position of dental enamel defects has become another standardised routine in archaeological studies, yielding much data which has been interpreted in a variety of ways. Bearing in mind the complex and nebulous picture drawn by modern epidemiological and dental research, the question of whether DDEs can realistically be used as a reliable indicator of health status must be asked. In this respect do we really know exactly what we are comparing? Are there related underlying causal relationships which inhibit enamel development or are these factors so diverse that interpretation cannot proceed beyond generalised statements, which may themselves not represent the true facts?

It was with these questions in mind that both authors commenced studies of DDEs on contemporary groups in order further to establish possible links between DDEs and such factors as general nutritional and health levels, social status, and ethnicity. The studies dealt with in this paper have contrasted wholly different populations: one from rural Mexico and the other from a large urban centre in the United Kingdom.

Materials and Methods

Two different population samples were studied. One is based in the rural community of Tezonteopan, central Mexico, and was designed to compare the effects of undernutrition on the occurrence of DDEs (Goodman *et al*, 1989). The other is a longer term study on the prevalence of DDEs in contrasting ethnic and socio-economic groups present in Bradford, West Yorkshire.

The methods of collecting these data were identical in practise and varied only in organisational procedures. The data from the Bradford study were collected in two distinct phases. Primary data was collected whilst accompanying the community dentists on regular school dental inspections, where at any one visit perhaps as many as 300 children would have been examined briefly. During this phase all types of DDEs, visible on the anterior dentition, were recorded. Those exhibiting specifically linear enamel hypoplasia (LEH) or pits, two categories of DDEs, were subsequently recalled to the dental clinic where more detailed recording of position and severity of each lesion was made.

The data presented from Bradford in this paper is from the first phase only. The clinical procedures undertaken during the Mexican study would equate to the second more rigorous stage.

In both studies the 12 anterior permanent teeth of each child were viewed with the aid of a small pencil torch. These were held at an oblique angle in order to highlight abnormalities on the surface topography of the tooth. Each defect was then recorded using a modified version of the criteria established by the Federation Dentaire International (FDI, 1982). This sets out clearly defined standards for the identification and systematic recording of DDEs. Thus the term DDE refers to any kind of enamel defect clinically identified using the FDI standards.

Five categories of defects could be reliably distinguished: demarcated opacities, diffuse opacities with fine lines, diffuse opacities without lines, pit hypoplasias and horizontal lines or grooves (LEH). For the purposes of this study, statistics are given on the total frequency of all DDEs, with the quantitative defects (LEH and Pits) dealt with separately. In other studies LEH and pits have been classified together (see Table 8.1). This can result in possibly confusing terminology which has varied from one researcher to the next. The F.D.I. index therefore provides a sound basis from which to work. The reason for the emphasis on LEH and pits is that these are the most commonly occurring DDEs in archaeological assemblages with lesions not restricted to human teeth but often present on the teeth of various mammals (Dobney, 1983).

As a way of testing visual recording repeatability, as well as gleaning information not easily visually recorded on posterior teeth, high quality dental impressions were taken where possible using a hydrophilic vinyl polysiloxan impression material. A low viscosity light body or "wash" was then applied which accurately replicated tooth surface detail. These ensured the collection of a permanent archive of dentition replicas, from which not only the visual recording methods could be checked, but detailed measurements of position and severity of lesions on each tooth could be taken at a later date.

a) Bradford

Sampling and supplementary information.

Ten schools were selected from seven different electoral wards in the Bradford Metropolitan region with a total of 2153 children viewed and included in this study. The areas involved were ideal since they included both socio-economically and ethnically diverse population samples.

All schools had a relatively local catchment area with the exception of St. Blaise Roman Catholic school which drew pupils from the environs of Leeds, some six miles away. Thus for the purpose of this study each school could be regarded as a relatively homogeneous group, although obvious movement between wards, other counties and even countries has occurred in a few instances.

Socio-economic criteria were calculated broadly on the basis of each ward using the Jarman 8 index. This is but one of a variety of socio-deprivation indices used regularly by clinical epidemiologists. The criteria generally used includes census data on pensioners alone, one parent families, the unskilled, unemployed, overcrowding, under fives, mobility and ethnic minorities (Jarman, 1983). This method provides a crude but composite socio-economic indicator . The Jarman 8 scores for Bradford are derived from a recent publication by the local health authority, "The Health of Bradford 1987" where indices are calculated for 21 of the electoral wards. The higher the Jarman 8 index the more socially underprivileged an area is deemed to be. Using this method the Bradford Metropolitan region falls tenth worst of the 192 districts of England whilst within the region scores range widely from -3.9 in the Baildon electoral ward to +72.3 around the University. The schools in this study are situated in wards where the Jarman 8 indices range from +10.00 to +72.3 (see Table 8.2).

Tezonteopan

Sampling and supplementary information

This project took place in Tezonteopan, a Nahuatl community located 180 km to the south of Mexico City in the central Mexican highlands. It is here that in the 1960s Chavez and Martinez began a nutritional supplementation study in order to understand better the effect of chronic undernutrition on functions such as disease resistance, growth and activity (Chavez and Martinez, 1982). Tezonteopan was originally selected for study because it was relatively isolated and "typically poor" (ibid), with inhabitants engaging in subsistence agriculture and eating a traditional maize and beans based diet. Shortly after the start of the study, a clinic was built to house the project staff and act as a base from which to conduct the study. It is here that the supplement was and still is distributed and consumed.

The basic design of the supplementation study involved the recruitment of two groups of mother-child pairs. The non-supplemented group was recruited in 1968. To be included in this group mothers had to be free of obvious diseases, have had between one and four pregnancies, be between 18 and 36 years of age, and be within one Z-score of the mean height and SES for the community. Newborns needed to have a birth weight of greater than 2.5 kg and an APGAR of eight or greater.

The same criteria were used for selection of the supplemented group. Great effort was made to match as closely as possible to the non-supplemented on a variety of criteria. In fact many of the matched infants are same-sex siblings (see Figure 8.2 in Goodman, this volume).

The nutritional supplement was given daily after the first missed menstruation. Supplements were given to the mother during pregnancy and lactation, and thereafter to the infant. The supplement contributes an average of 300 extra calories and 20 extra grams of protein to the daily diet.

Prior studies have shown that the nutritional supplement had a profound effect on such variables as growth, activity levels and percent of days ill. Differences in growth between groups are profound. Height differed by three inches at age two and later reached nearly five inches. The non-supplemented children suffered nearly double the number of days ill with respiratory and diarrhoeal illness. In sum, the supplement had a profound effect on nutritional status and has been isolated as the main variable differentiating the two groups.

All study participants who were ten years of age or older were examined. Only a couple of participants had left Tezonteopan since the start of the study. The sample consisted of forty two pairs, equally divided between males and females. The recorder was never informed as to the supplemented status of the child, that is, this was a blind study.

Results

The following are preliminary results obtained on the frequencies of DDEs from these two study samples.

a) Bradford

Prevalence of DDEs by Electoral Ward

An initial broad comparison can be made between each school according to the frequency of children exhibiting at least one type of enamel defect. Thus it can be seen from Table 8.2 that these prevalence figures range from 42% with at least one enamel defect at Priestman middle school to 83% at Whetley middle school. Working on the premise that health and social deprivation are intimately linked, and that the frequency of DDEs is a reliable index of these factors, it should follow that frequencies of DDEs increase in more socially deprived areas. In other words, where the Jarman 8 index is high, the frequency of DDEs should be similarly so. This, however, is not the case in Bradford.

The percentage of children exhibiting DDEs for each school ranked by their Jarman 8 indices is shown in Figure 8.1. Thus Mandale middle school on the far left has the lowest Jarman 8 score, whilst Waverley middle school on the far right has the highest. If the above assumption is true then we would expect a general rise in the frequency of DDEs from left to right. This is clearly not the case. In fact, two of the schools with the highest Jarman 8 scores, Ryan and Waverley, have the second and third lowest occurrences of DDEs. Conversely Great Horton middle school, whilst having the third lowest Jarman 8 score, exhibits the second highest frequency of DDEs. However, of the four schools which share almost identical Jarman 8 scores, the range of frequencies is close indeed, from 59% to 70%. In fact, three of them differ by as little as 3%.

Broadly similar patterns emerge when considering the frequency of LEH present in each of the schools (Figure 8.2). Again Waverley and Ryan middle schools exhibit the second and third lowest frequency of LEH, whilst Mandale and Gt. Horton middle schools equally show the second highest. Whetley middle school, with a Jarman 8 index of 44.9, has the fourth lowest frequency of LEH (12%) while having the highest frequency of all types of DDEs. In the case of LEH, the four schools with almost identical Jarman 8 scores show frequencies ranging from 24% to 41% with the highest frequency of all occurring at St. Blaise Roman Catholic school. Interestingly, as described earlier, this was the only school where pupils were obtained from a much wider and less cohesive catchment area. The lowest frequency of LEH (7%) is found at Priestman middle school which similarly has the second lowest Jarman 8 score of 17.4.

Tooth prevalence of linear enamel hypoplasia

As well as contrasting frequencies of LEH between individuals (mouth prevalences), frequencies were also studied between teeth (tooth prevalences). Previous research has indicated that certain teeth appear to be more susceptible to enamel disturbances than others. As a result, mouth frequencies may differ significantly depending on which teeth are available for study (Goodman and Armelagos, 1985). Figure 8.3 shows the tooth prevalence of LEH contrasted between two schools with significantly different Jarman 8 scores, these being Mandale and Undercliffe middle school. Thus the likelihood of LEH occurring on the permanent maxillary central incisors appears to be *almost doubled* for those children attending Mandale whilst quite the reverse is ture of the canine teeth, although numbers in this case are much reduced. There seems to be a much closer correspondence between schools when the mandibular permanent dentition is considered although differences between LEH present on the canine teeth remain the same.

Although there are significant differences between these two schools, the overall differences between each tooth type are consistent with prior studies (Goodman and Armelagos, 1985). This showed that highest frequencies were to be found on the central incisors, followed by the lateral incisors and lastly the canines.

Ethnic group differences

Socio-economic factors are not the only obvious controlling variables involved in this sample. In Bradford the picture is made more complex by an ethnically diverse community, including Asians, Bengalis and Indians, a majority of Pakistanis, as well as English Europeans. Two schools with differing Jarman 8 scores, and where both European and Asian children are present in comparable numbers, were again compared for the occurrence of DDEs. These were Gt Horton and Undercliffe middle schools.

On the basis of mouth prevalence of DDEs alone, it is clear that there are marked differences between the two schools (see Figure 8.4). At Undercliffe there are 20% more European children showing DDEs than at Gt. Horton, what one would expect when also considering the Jarman 8 index. The situation is, however, reversed when considering the Asian children where 25% more DDEs are present at Gt. Horton. When contrasting the ethnic groups within schools a striking difference is obvious; whereas at Undercliffe frequencies of DDEs are almost identical for each group, at Great Horton there is a 44% difference suggesting that Asian children in this locality are almost twice as likely to be susceptible to disturbances in enamel formation than European children. This again contrasts strongly with the socio-deprivation indices.

Turning to the occurrence of LEH and pits it appears that, although with a lower Jarman 8 score, these conditions are far more common in the Gt. Horton sample (Figure 8.5). The statistics indicate that there is little difference between the ethnic groups at Undercliffe with regard to the occurrence of both LEH and pits but that at Gt. Horton pits are marginally more common in the Asian sample.

Since combinations of DDEs were often seen to be present on and between teeth it was thought that perhaps mouth prevalences of the total numbers of conditions would vary between samples (Figure 8.6). The presence of only one type of DDE is more common at Undercliffe middle school than at Gt. Horton. However, whereas it is more common for one condition to be present in Europeans at Undercliffe, the situation is reversed at Gt. Horton, albeit marginally. Two conditions are more likely to occur at Gt. Horton but frequencies are similar between the two ethnic groups. Three conditions are not as numerous, being slightly more common in Asians than Europeans at Undercliffe and similar for both groups at Gt. Horton. A combination of four conditions is less likely and in this sample is only present in the Asian sample although small in number.

b) Tezonteopan

The tooth specific prevalence of pits is presented in Figures 8.7a and 8.7b for the supplemented and non-supplemented groups. Overall frequencies of pits by tooth are low and variable, ranging from 0% to 11.9% prevalence. In most cases, and especially for the upper left lateral incisor and most of the lower teeth, there is a greater but not significantly different frequency of pits in the supplemented group. There is *no evidence* to suggest that the supplement protected against the formation of these defects. A possible explanation for these defects may be that they are due to local trauma. This is supported by their irregular pattern and the fact that the supplemented group were much more active and exploratory as infants and children (Chavez and Martinez, 1982).

In contrast to the pit defects, the frequency of LEH by maxillary and mandibular anterior tooth is higher and significantly reduced in the supplemented group (Figures 8.8a and 8.8b). Consistent with prior studies, the highest frequency of defects are found on the central incisors. For the maxillary teeth, the frequency of LEH by tooth ranges from 2.4% (maxillary left lateral incisor) to 40.5% (maxillary left central incisor) in the supplemented group, while the range is from 15% (left upper canine) to 71.4% (right upper central incisor) for the non-supplemented group. Differences are significant at the .05 level for all maxillary left sided teeth and the right central incisor ($p<.002$). The relative risk of an LEH is nearly double for the non-supplemented versus the supplemented group.

The chronological pattern of LEH for the supplemented and non-supplemented groups has also been constructed based on the developmental standards of Massler and co-workers (1941) (Figure 8.9). The distributions by developmental ages, based on upper central incisor data, are relatively similar with a "peak" between two and three years developmental age. This is precisely the age at which infection and diarrhoeal disease peaks in this community and when breastfeeding ends (Chavez and Martinex, 1982). The greatest difference in the two patterns is between birth and one year and three and four years, or before and after weaning. It is as if *all children* are at risk during weaning, but the supplemented are *protected* before and after weaning.

Discussion

The data presented yield a complex and seemingly confliciting picture. The results of the Bradford study suggest that the variety of frequencies of DDEs does not in general correspond to socio-economic criteria, which in turn is intimately linked with health status. In fact, in the Bradford sample quite the reverse seems to be the case where a higher socio-economic status appears to render an individual more susceptible to developmental disturbances of the enamel. it may be that the data are highlighting weaknesses in the criteria used to calculate the socio-deprivation indices, but in general this would seem unlikely, working on the premise that factors affecting soci-economic status are better understood than those which give rise to DDEs.

In contrast the Tezonteopan data indicates clear correlations between levels of nutrition and LEH. The results suggest that the non-supplemented individuals are generally twice as likely to exhibit linear defects than supplemented ones. In both cases a number of variables, unavailable in archaeological assemblages and which have been shown by research to be associated with disturbances of the enamel, are identified. Thus if the Bradford samples were archaeological assemblages, assuming the socio-economic and ethnic associations of DDEs to be correct, wholly false conclusions would be drawn. The opposite would of course be true for the Tezonteopan data.

What factors might explain these apparently conflicting results? Perhaps the most basic one lies in the fact that one sample derives from a developed country whilst the other is from a developing one. It follows that the frequency and magnitude of "stressors" should be greater in the Mexican study, due to marked contrasts in general living conditions between Tezonteopan and Bradford. However, it must be remembered that even within so-called developed countries a proportion of the population may experience a very low standard of living, perhaps not as marked as in parts of the third world, but nevertheless not wholly distanced from them. In fact, the majority of the individuals included in the Bradford study derive from these lower tiers of the social spectrum.

It is therefore likely that at least some of the factors involved in the production of DDEs are perhaps constant in samples from both undeveloped areas and low socio-economic groups from developed countries. When dealing with data from the developed world it is important, then, to distinguish socio-economic criteria quite clearly.

Socio-economic status

This is a highly relevant factor intimately linked with health status and a number of other variables. Previous studies have contrasted DDEs and socio-economic factors, with mixed results.

Enwonwu (1973), during research in Nigeria, found definite correlations between DDEs and socio-economic status. His results showed that DDEs were singularly more common in communities of very low socio-economic status and conversely almost totally absent from those at a much higher level. His groups, however, did represent absolute opposite ends of a much wider social spectrum than those generally encountered in more developed countries.

Dummer *et al* (1986a, 1986b), during studies in South Wales, found high prevalence of DDEs but fairly low levels of LEH. However, there were no significant differences between children from social classes i and iv and v in mouth prevalences of DDEs.

Smith (1983) suggested a correlation between DDEs and social status by showing a higher incidence of enamel hypoplasia in children living in council houses in the London borough of Havering.

Nation *et al* (1987) found no such differences and in fact found lower frequencies of DDEs in the primary teeth of Hispanic children compared with their Caucasian counterparts in California, even though many Hispanic families were of a lower educational status.

The Bradford data, as already stated, suggest no real correlation between low socio-economic status and increased frequencies of general DDEs or more specifically LEH. On the basis of the Jarman 8 index the data generally suggest a higher prevalence of DDEs in the least socially deprived wards, and a lower prevalence in those most socially deprived.

Nutritional status

General health goes hand in hand with social deprivation and one of the most important factors related to growth and development is nutrition. As well as arresting basic metabolic processes, malnutrition predisposes individuals to a variety of opportunistic infectious agents. A number of workers have therefore attempted to study the relationship of malnutrition to DDEs.

Infante (1974) recorded high prevalences of enamel hypoplasias on the teeth of children from an Apache Indian reservation who were suffering from acute diarrhoeal problems. He concluded that enamel hypoplasias could therefore be related to nutritional imbalance.

More directly, Sweeney *et al* (1971) and Infante and Gillespie (1974), working in Guatemala, similarly showed correlations between malnutriton and enamel hypoplasia. Sweeney's sample exhibited a substantially high prevalence of defects; 73% suffering from third degree malnutrition showed enamel hypoplasia whilst only 42% of those suffering from second degree malnutrition showed similar lesions.

Sawyer and Kwoku (1985) showed that in the deciduous teeth of a small sample of Nigerian children, only 15% of malnourished children showed LEH, whilst lesions were totally absent in those deemed well nourished.

Goodman *et al* (1987) also found positive correlations between nutritional status and the presence of DDEs in an earlier Mexican study from the Solas valley. The data from Tezonteopan also supports a nutritional link with frequency of DDEs, with clear correlations existing between LEH and chronic low grade malnutrition. With the introduction of a dietary supplementation the frequency of LEH is significantly reduced.

Ethnic Differences

In a number of other studies where contrasts were specifically made between different ethnic groups there appeared to be substantial variation in the frequencies of DDEs recorded.

During the study of DDEs in a variety of temperature and fluoride zones of North and Central America, Richards *et al* found little significant difference between Caucasians and non-Caucasians. King and Wei (1986), however, found clear differences between Chinese/Malay, Indian and Caucasian children studied in Hong Kong and Singapore. Here the Chinese/Malay communities were worst affected followed by Caucasians, with Indians being least affected. Schamschula *et al* (1980) report the frequencies of DDEs in an Australian Aboriginal sample to be almost twice as common as a similar Caucasian one. However, the overall frequencies noted are much lower than from many other studies. Lunz (1987), during work in South Africa, also found enamel hypoplasias most common in the teeth of children from Chinese families. These were followed closely by Black children and finally by Caucasians who showed almost negligible amounts of enamel hypoplasia. Differences within the Black sample were also noted where significant. More hypoplasias were found in tribal groups that were mainly pastoralists, and less in those that were classified as hunter-gatherers or mixed pastoralists.

It seems clear that ethnic differences of DDE frequencies are often inextricably linked with socio-economic status. In many instances ethnic groups showing high frequencies of DDEs are those very ones which are most socially disadvantaged.

From preliminary results of the Bradford study it would seem that DDEs are in general more frequent in Asian than in European children. There are also differences within these groups between the varying schools suggesting that further underlying complicating factors are probably involved.

The Bradford data fits well with previous studies in developed parts of the world where few associations have been found between DDEs and a number of other criteria. In contrast, the Tezonteopan data supports more recent studies in underdeveloped regions where correlations have been made with nutritional, socio-economic and ethnic factors. This suggests that perhaps studies in developed countries of the world are inherently confused by the presence of what can be termed "background noise".

It is clear from modern clinical research that there may be over 100 environmental causative factors which can result in the production of a DDE (Cuttress and Suckling, 1982). This supports Kreshover's statement of almost 30 years ago that "ample clinical and experimental evidence exists to suggest that tooth defects are generally non-specific in nature and can be related to a wide variety of systemic disturbances any of which, depending upon their severity and degree of tissue response, might result in defective enamel" (1960:166).

It is possible that as one ascends the socio-economic ladder the "noise to signal" ratio increases so that more fundamental signals, in the form of major stressors, are increasingly masked by other less important ones. This "noise", although still capable of interfering with enamel formation, may perhaps be of secondary importance to core factors such as nutrition which appears to play a major role during childhood development and adaptation, and ultimately in the production of DDEs in developing regions. That is not to say, however, that frequencies and patterns of DDEs in developed countries are wholly random, but may in fact have a more diffuse basis.

General factors involved with the production of DDEs in archaeological assemblages are more likely to correlate closely with those found in present day rural undeveloped communities, like Tezonteopan, than more developed ones. Comparisons of epidemiological data from modern undeveloped communities with ancient skeletal assemblages are therefore more likely to be valid.

Physical anthropologists have assumed for some time that DDEs can be considered as general stress indicators in archaeological assemblages. Is this really possible? In many of the modern studies some DDEs seem to appear quite randomly (Smith, 1983). Newton *et al* found chronological associations between only 50% of the teeth removed from a sample of 76 still born and infant deaths. Suckling *et al* (1987) found few positive associations between DDEs and possible aetiological factors in a group of New Zealand children. This is not surprising since information on nutrition and other important factors was often not available.

Many LEH lesions are subtle indeed, often no more than a series of exaggerated perikymata[1]. Histological research undertaken on these defects shows few changes corresponding with the surface lesion (Goodman and Rose, in press). So should these be regarded as pathological or merely manifestations of a normal range of enamel? Perhaps many of the differences of frequencies recorded in previous studies, and likewise between Tezonteopan and Bradford, could well be the result of direct differences in the magnitude of observations between researchers. This could apply equally well to data from archaeological studies (Goodman *et al*, 1980).

There is no doubt that DDEs, particularly LEH and pits, commonly occur on both modern and archaeological teeth. But are comparisons of the frequency of hypoplasias within and between populations possible? Ignoring for the moment the daunting variety of known and unknown stimulus events which can result in the production of DDEs, let us consider the premise that their presence can tell us something about the chronology of non-specific stress. Here we must consider the concept of a critical point or threshold.

It is obvious that a DDE will not occur until specific stimulus events compromise the healthy individual sufficient to reach and cross this critical threshold. At this point, ameloblast activity may be directly or indirectly affected, depending on the severity and duration of these stimulus events. It is assumed that the more severe and protracted this stimulus the greater the impact on the enamel and thus the larger, more frequent and more severe the lesion will be. Can we also assume that the same stimulus event will affect all individuals equally?

Figure 8.10 shows a simplistic hypothetical model of three archaeological/modern populations (they could equally represent three individuals within a population). The vertical axis represents health status, which here becomes progressively worse as once ascends, and the horizontal axis represents chronological time during tooth development. The dotted line represents the critical point or threshold which must be crossed in order to produce a DDE.

From this simple model it is clear that with populations all of differing health status, suffering exactly the same number and magnitude of stimulus events, the resulting pattern of DDEs on the teeth will be wholly different. It does, however, accept that the closer the population/sample/individual is to that threshold the more likely the stimulus events are to be manifested as enamel disturbances. Thus more highly compromised groups should in theory exhibit more DDEs. The data from Tezonteopan supports this hypothesis where a lower nutritional status results in a larger proportion of defects being caused by the signal illness or nutritional stress.

The clinical picture is possibly further confused by the fact that the critical threshold and health status of populations and individuals are not static variables and so may be too simplistic to represent them on the model as merely straight lines. Surely in a living population, where complex relationships exist between the individual and the environment, the threshold would be quite dynamic.

There are obvious problems and limitations when considering patterns of DDEs within and between living populations even where many of the variables which indicate general health status are available. These limitations are compounded in archaeological assemblages where the information is totally absent or, where present, is circumstantial. Where hypoplasias are common in an assemblage it is clear that stimulus events

---

[1] Perikymata are the expression on the tooth surface of Striae of Retzius, the lines which represent the developing margin of the enamel.

are occurring at particular times during tooth development. Thus the position of lesions on the tooth crown is extremely important since chronological sequences of stimuli can be assessed. This approach may also be somewhat problematic since the criteria established for the chronology of tooth development derives from individuals from a developed society, perhaps not readily applicable to modern rural undeveloped or archaeological samples.

Conclusions drawn from comparisons between ancient populations must therefore be guarded unless the skeletal assemblages can be shown, through archaeological and possibly historical research, to be comparable i.e. that the environmental and social factors relevant to the production of DDEs are similar.

Conclusions

For too long physical anthropologists have used enamel hypoplasias as a simplified and wholly standardised method of assessing the overall health status of skeletal assemblages without a clear appreciation of the complexities still being unravelled by modern epidemiological research. It is clear that some DDEs may seem random in their appearance i.e. involving as yet unknown aetiological factors. Some lesions may indeed not even be pathological. Identical frequencies and chronological patterns of DDEs in two separate populations may, in reality, reflect wholly different variables involved in their production. The health status of individuals within and between populations may vary significantly with wholly different patterns of DDEs resulting from ostensibly similar stimulus events.

Modern clinical, histological and epidemiological research is beginning to unravel some of the complex factors involved in the producition of DDEs but there are still many questions which remain unaswered.

References

Beynon, A.D. and Wood, B.A. (1987). Patterns and rates of enamel growth in the molar teeth of early hominids. *Nature 326*:493-96.

Bromage, T. and Dean, M.C. (1985). Re-evaluation of the age at death of immature fossil hominids. *Nature 317*:525-27.

Chavez, A. and Martinez, C. (1982). *Growing Up in a Developing Country*. Mexico: Instituto Nacional de la Nutricion.

Chelloppah, N.K., Lo, G.L. & Vignehsa, H. (1986). Enamel defects in permanent maxillary incisors of Singaporean children. *Ann. Acad. Med. Singapore* 5:280-83.

Cuttress, T.W. and Suckling, G.W. (1982). The assessment of non-carious defects of enamel. *Int. Dent. J. 32*:117-22.

Dobney, K. (1983). *Enamel Hypoplasia of Pig Teeth: Its Relevance and Application to Zooarchaeological Studies*. M.Sc. Thesis, University of London.

Dummer, P.M., Kingdon, A. & Kingdon, R. (1986a). Prevalence of enamel developmental defects in a group of 11 and 12 year old children in South Wales. *Comm. Dent. Oral Epidemiol. 14*:119-22.

Dummer, P.M., Kingdon, A. & Kingdon, R. (1986b). Distribution of developmental defects of tooth enamel by tooth type in 11-12 year old children in South Wales. *Comm. Dent. Oral Epidemiol. 14*:341-44.

Enwonwu, C.O. (1973). Influence of socio-economic conditions on dental development in Nigerian children. *Arch. Oral. Biol. 18*:95-107.

F.D.I. Commission on oral health, research and epidemiology (1982). An epidemiological index of developmental defects of dental enamel (DDE Index). *Tech. rep. no. 15 Int. Dent. J. 32*:159-67.

Goodman, A.H. and Armelagos, G.J. (1985). Factors affecting the distribution of enamel hypoplasias within the human permanent dentition. *Am. J. Phys. Anthropol. 68*:479-93.

Goodman, A.H. and Armelagos, G.J. (1988). Childhood stress and decreased longevity in a prehistoric population. *Am. Anthrop. 90*:936-44.

Goodman, A.H. and Rose, J.C. (in press). Dental enamel hypoplasias as indicators of nutritional status. In M.A. Kelly and C.S. Larsen (eds.) *Advances in Dental Anthropology*. Orlando: Academic Press.

Goodman, A.H., Armelagos, G.J. & Rose, J.C. (1980). Enamel hypoplasias as indicators of stress in three prehistoric populations from Illinois. *Hum. Biol. 52*:515-28.

Goodman, A.H., Allen, L.H., Hernandez, G.P., Amador, A., Arriola, L.V., Chavez, A. & Pelto, G.H. (1987). Prevalence and age at development of enamel hypoplasias in Mexican children. *Am. J. Phys. Anthropol. 72*:7-19.

Goodman, A.H., Martin, D.L., Perry, A., Dobney, K., Martinez, C., & Chavez, A. (1989). The effect of nutritional supplementation on permanent tooth development and morphology (abstract). *Am. J. Phys. Anthropol. 78*:229.

Gray, A.S. (1964). Prevalence of opacities and hypoplasias of nonfluoride origin in permanent teeth of children. *Can. Dent. Assoc. 30*:137-39.

Infante, P.F. (1974). Enamel hypoplasia in Apache Indian childhood. *Ecol. Food Nutr. 2*:155-56.

Infante, P.F. and Gillespie, G.M. (1974). An epidemiologic study of linear enamel hypoplasia of deciduous anterior teeth in Guatemalen children. *Archs. Oral Biol. 19*:1055-61.

Jarman, B. (1983). Identification of underprivileged areas. *Brit. Med. J. 286*:1705-09.

Jelliffe, D.B. and Jelliffe, E.F.P. (1961). The nutritional status of Haitian children. *Acta. Trop. 18*:1.

Jelliffe, D.B., Jelliffe, E.F.P., Garcia, I. & De Barrios, G. (1961). The children of the San Blas Indians of Panama. *J. Pediat. 59*:271-85.

Jontell, M. and Linde, A. (1986). Nutritional aspects of tooth formation. *World Rev. Nutr. Diet. 48*:114-36.

King, N.M. and Wei, S.H. (1986). Developmental defects of enamel: a study of 12-year olds in Hong Kong. *J. Am. Dent. Assoc. 112*:835-39.

Kreshover, S.J. (1960). Metabolic disturbances in tooth formation. *Ann. N.Y. Acad. Sci. 85*:161-7.

Lunz, R. (1987). Incidence and age of development of enamel hypoplasia in southern African human populations (abstract). *S. Afr. J. Sci. 83*:379.

Massler, M., Schour, I. & Poncher, H.G. (1941). Developmental pattern of the child as reflected in the calcification pattern of the teeth. *Am. J. Dis. Child.* *62*:33-67.

Menezes, D.M. (1976). Opacities and hypoplasia in the enamel of Burmese children from a low fluoride area. *J. Dent.* *4*:71-2.

Nation, W.A., Matsson, L. & Peterson, J.E. (1987). Developmental enamel defects of the primary dentition in a group of Californian school children. *J. Dent. Child.* *4*:330-4.

Newton, R.W., Levine, R.S., Turner, E.P. & Baron, A.J. (1984). The sensitivity of the developing tooth germ to systemic disturbances. *Early Hum. Dev.* *9*:269-73.

Richards, L.F., Westmoreland, W.W., Tashiro, M. & Morrisson, J.T. (1967). Nonfluoride enamel hypoplasia in varying fluoride temperature zones. *J. Am. Dent. Assoc.* *75*:1412-18.

Sarnat, B.G. and Schour, I. (1941). Enamel hypoplasia (chronic enamel aplasia) in relationship to systemic disease: a chronologic, morphologic and etiologic classification. *J. Am. Dent. Assoc.* *28*:1989-2000.

Sawyer, D.R. and Kwoku, A.L. (1985). Malnutrition and oral health of children in Ogbomosho, Nigeria. *J. Dent. Child.* *52*:141-45.

Schamschula, R.G., Cooper, M.H., Adkins, B.L., Barnes, D.E. & Agus, H.M. (1980). Oral conditions in Australian children of Aboriginal and Caucasian descent. *Comm. Dent. Oral Epidemiol.* *8*:365-69.

Smith, J.M. (1983). *Epidemiology, Aetiology and Histopathology of Developmental Enamel Defects in Human Teeth.* Ph.D. Thesis, University of London (London Hospital Medical School).

Suckling, G.W. and Pearce, E.I.F. (1984). Developmental defects of enamel in a group of New Zealand children: their prevalence and some associated etiological factors. *Comm. Dent. Oral Epidemiol.* *12*:177-84.

Suckling, G.W., Herbison, G.P. & Brown, R.H. (1987). Etiological factors influencing the prevalence of developmental defects of dental enamel in nine-year-old New Zealand children participating in a health and development study. *J. Dent. Res.* *66*:1466-69.

Sweeney, E.A., Saffir, A.J. & de Leon, R. (1971). Linear hypoplasia of deciduous incisor teeth in malnourished children. *Am. J. Clin. Nutr.* *24*:29-31.

The Health of Bradford (1987). Bradford Health Authority Publication.

Acknowledgements

The authors are grateful to both the National Institute of Health (USA) (Grant No. DE08607) and the Science and Engineering Research Council (UK) for generously funding the studies. Thanks are also due to the 3M company for providing the impression materials used in Tezonteopan.

We are indebted to the Mexican National Institute of Nutrition and Bradford Health Authority for their co-operation during the research. Special thanks are due to the community dental officers and nurses in Bradford and to the parents and children who participated in both studies.

TABLE 8.1

SELECTED PREVIOUS STUDIES OF DDEs

LOCATION (Author)	YEAR	SOCIO-ECONOMIC ETHNIC	DDE	LEH	PITS	COMMENTS
USA and Central AM (Richards)	1967	Caucasian non Caucasian	38% 34%			
Australia (Schamschula)	1980	Aboriginal Caucasian	17% 8%			
Hong Kong (King)	1986	Chinese Caucasian Indian			79% 56%	Summarised hypoplasias Both LEH and Pits
Singapore (Chellopah)	1986	Indians Chinese/Malay Total	56% 74% 72%			Based on Enamel Opacities on upper Incisors only
South Africa (Lunz)	1987	Caucasian Chinese Black Total		3% 59% 37% 27%		
California (Nation)	1987	Caucasian Hispanic Black Asian	38% 18% 60% 0%		23% 10% 50% 0%	Based on Primary teeth Summarised hypoplasia Both LEH and Pits
Haiti (Jelliffe)	1954			14%		Based on Primary teeth
Panama (Jelliffe)	1961			31%		Based on Primary teeth
Canada (Gray)	1964		14%			
Burma (Menenez)	1976		52%	30%		Both LEH and Pits
London (Smith)	1983		66%	15%		
New Zealand (Suckling)	1984		63%	3%	9%	
South Wales (Dummer)	1986		49%		7%	Both LEH and Pits
Guatemala (Sweeney)	1971	3rd Deg Maln 2nd Deg Maln		73% 42%		Based on Primary teeth
Nigeria (Sawyer)	1985	Malnourished Well nourished		15% 0%		Based on Primary teeth
Nigeria (Enwonwu)	1973	Village sample Optimal sample			21% 0%	Both LEH and Pits

DDE = total of all varieties of developmental defects of enamel
LEH = linear enamel hypoplasia
Pits = single or multiple enamel pits

TABLE 8.2

FREQUENCY OF DENTAL ENAMEL DEFECTS IN BRADFORD BY SCHOOL AND WARD

SCHOOL	ELECTORAL WARD	JARMAN 8 INDEX	TOTAL NO. CHILDREN	DDE%	LEH%	PIT%
Waverley	University	72.3	307	46	8	8
Ryan	Little Horton	61.4	214	48	11	8
Whetley	Toller	44.9	290	83	12	6
Undercliffe	Undercliffe	39.3	151	67	24	11
Holme Wood	Tong	38.8	147	67	25	9
Tyersall	Tong	38.8	106	59	31	21
St Blaise	Tong	38.8	326	70	41	8
Gt Horton	Gt Horton	18.8	132	74	36	22
Priestman	Wibsey	17.4	259	42	7	7
Mandale	Queensbury	10.0	221	66	36	12
		TOTAL	2153	-	-	-
		MEAN	-	62%	23%	11%

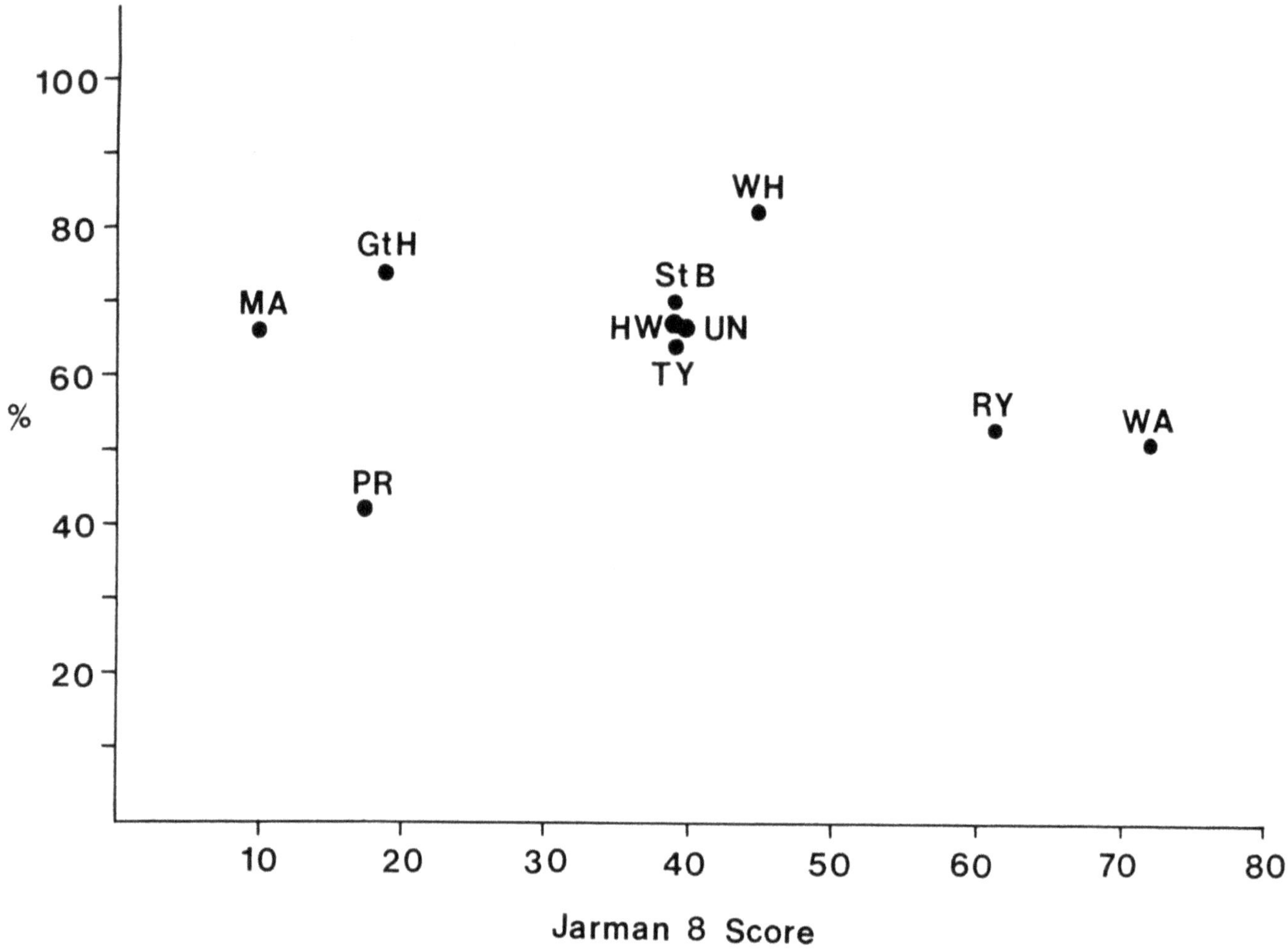

Figure 8.1 Frequency of DDEs from children examined at ten schools in the Bradford region. Jarman 8 scores indicate the socio-economic status of each.

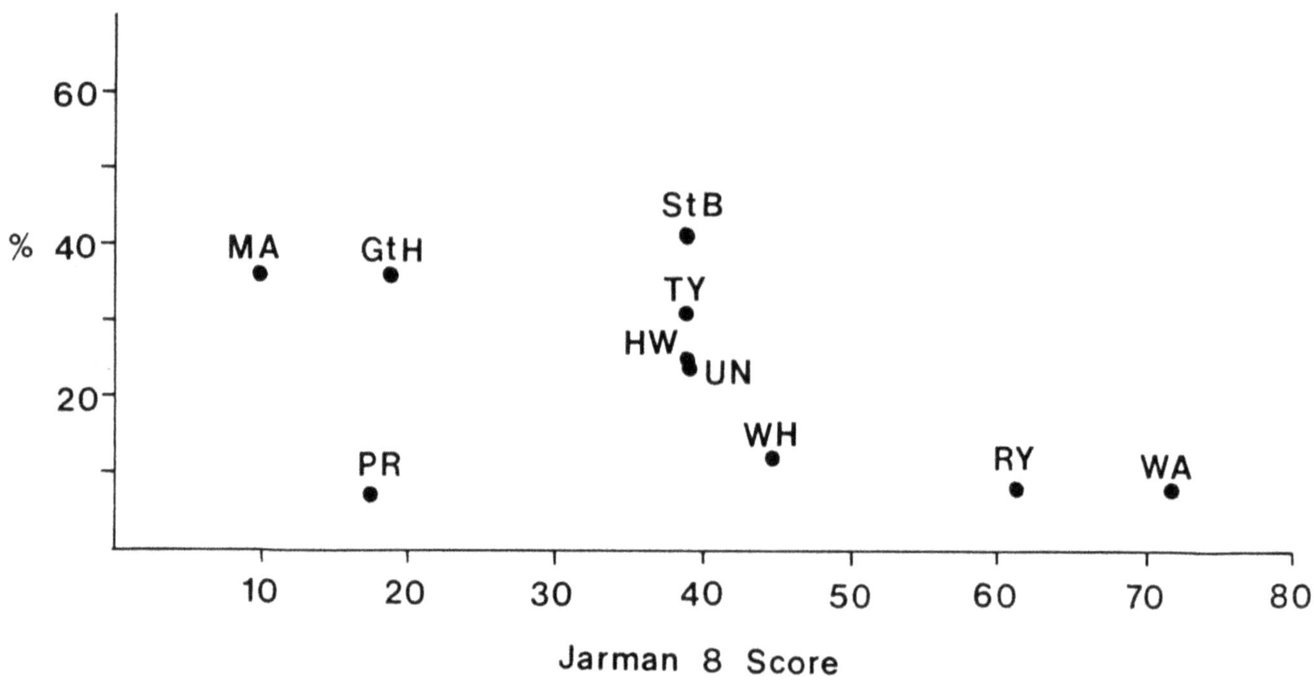

Figure 8.2 Frequency of LEH from children examined at ten schools in the Bradford region. Jarman 8 scores indicate the socio-economic status of each.

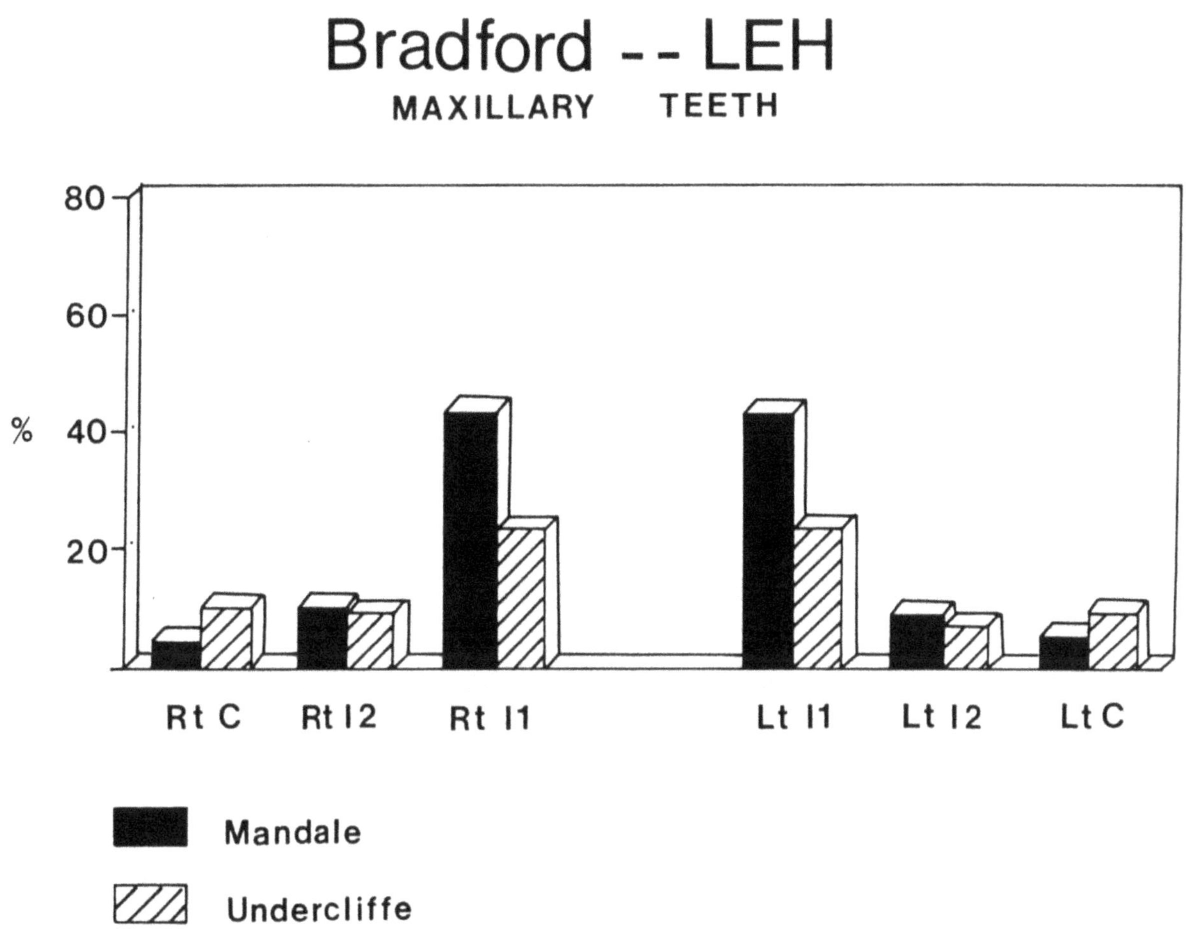

Figure 8.3 Bradford LEH, maxillary teeth.

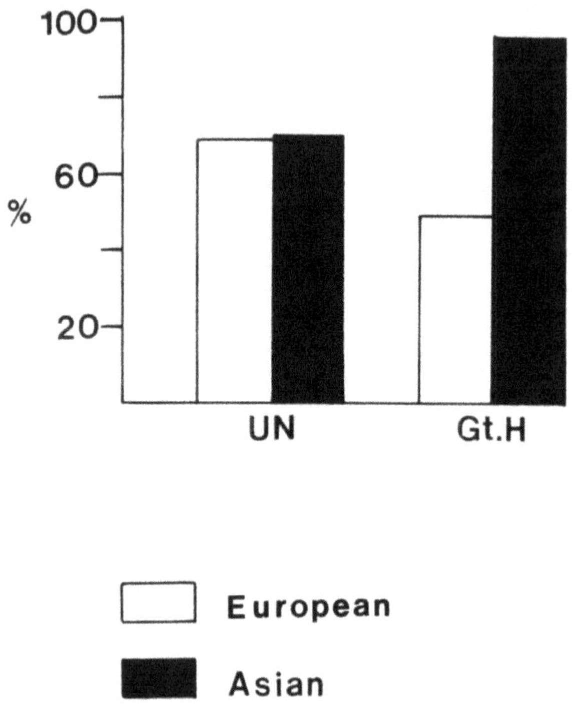

Figure 8.4 Frequency of DDEs in European and Asian children from Undercliffe and Gt. Horton middle schools.

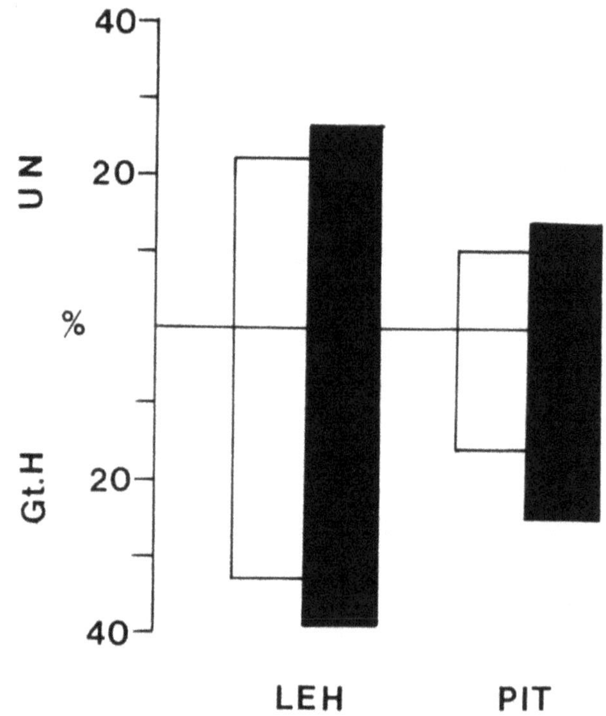

Figure 8.5 Frequency of LEH and pits in Asian children from Undercliffe and Gt. Horton middle schools.

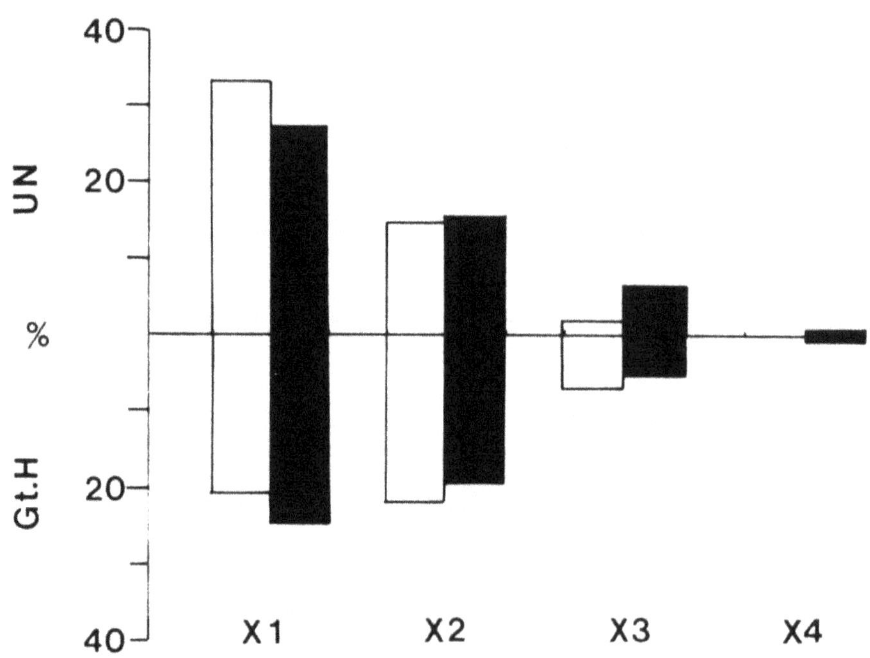

Figure 8.6 Frequency of different types of enamel defects in European and Asian children from Undercliffe and Gt. Horton middle schools.

Tezonteopan -- Pit HP
Supplemented vs Non-Supplemented

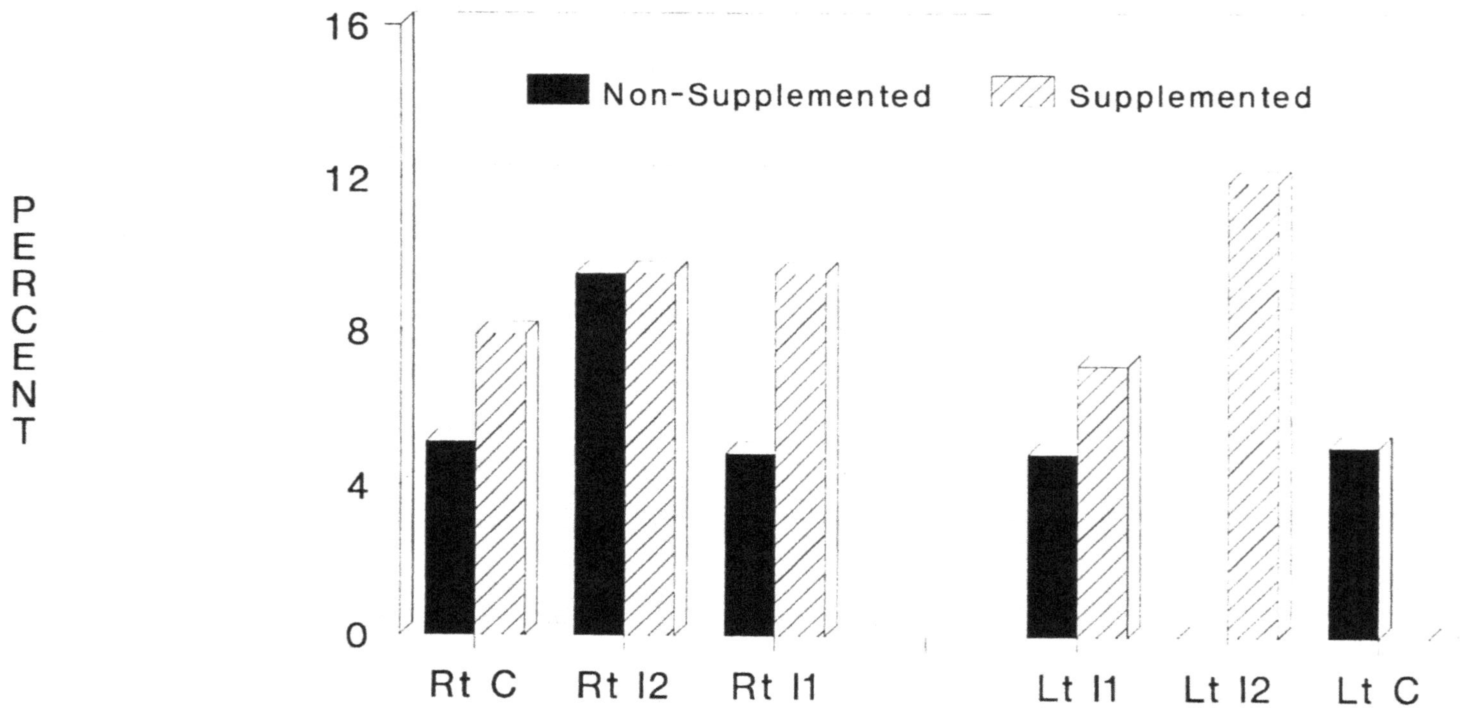

	Rt C	Rt I2	Rt I1	Lt I1	Lt I2	Lt C
Non-Supplemented	5.1	9.5	4.8	4.8	0	5
Supplemented	7.9	9.5	9.5	7.1	11.9	0

Figure 8.7a

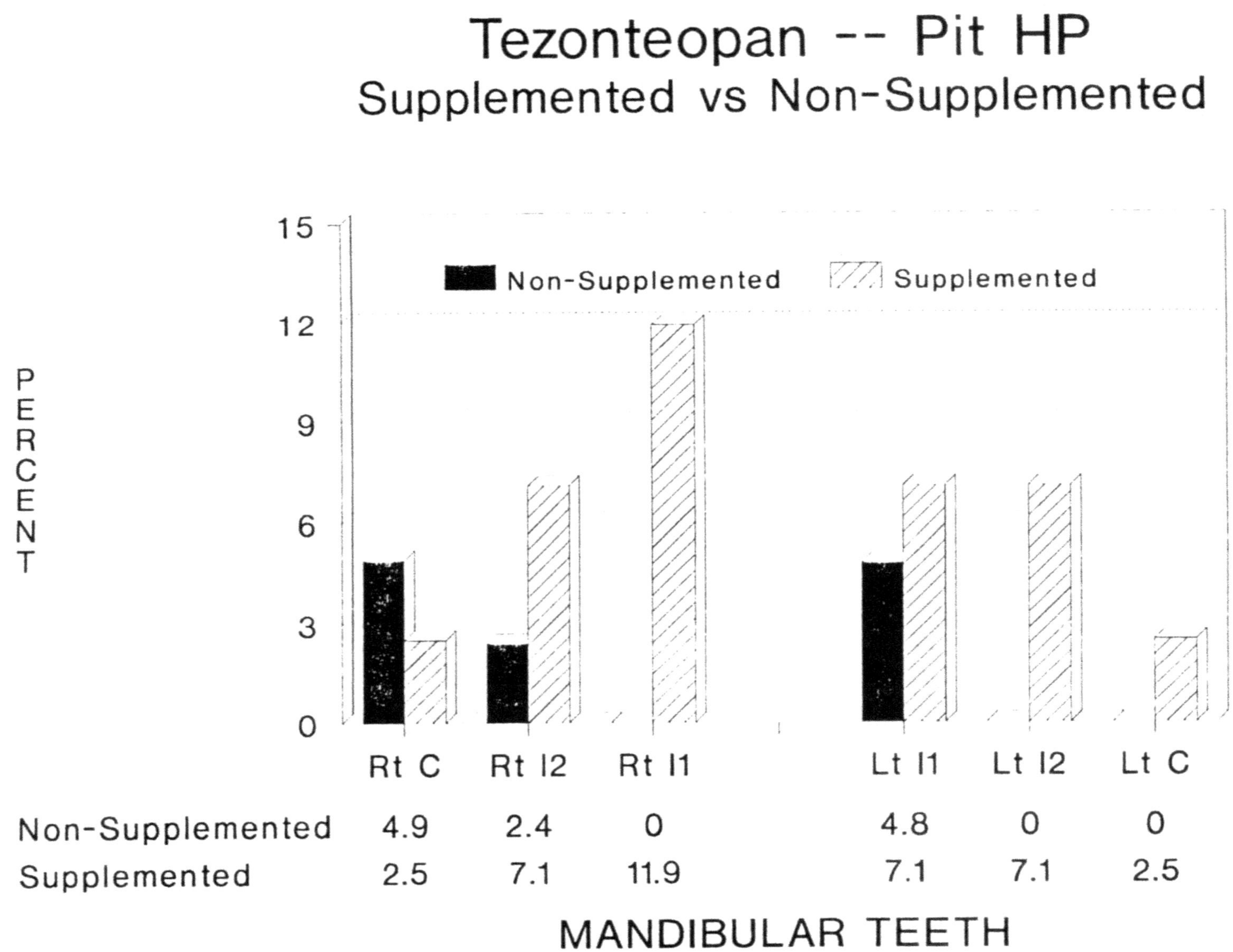
Tezonteopan -- Pit HP
Supplemented vs Non-Supplemented
Non-Supplemented
Supplemented
15
12
9
6
3
0
PERCENT
Rt C
Rt I2
Rt I1
Lt I1
Lt I2
Lt C
Non-Supplemented 4.9 2.4 0 4.8 0 0
Supplemented 2.5 7.1 11.9 7.1 7.1 2.5
MANDIBULAR TEETH

Figure 8.7b

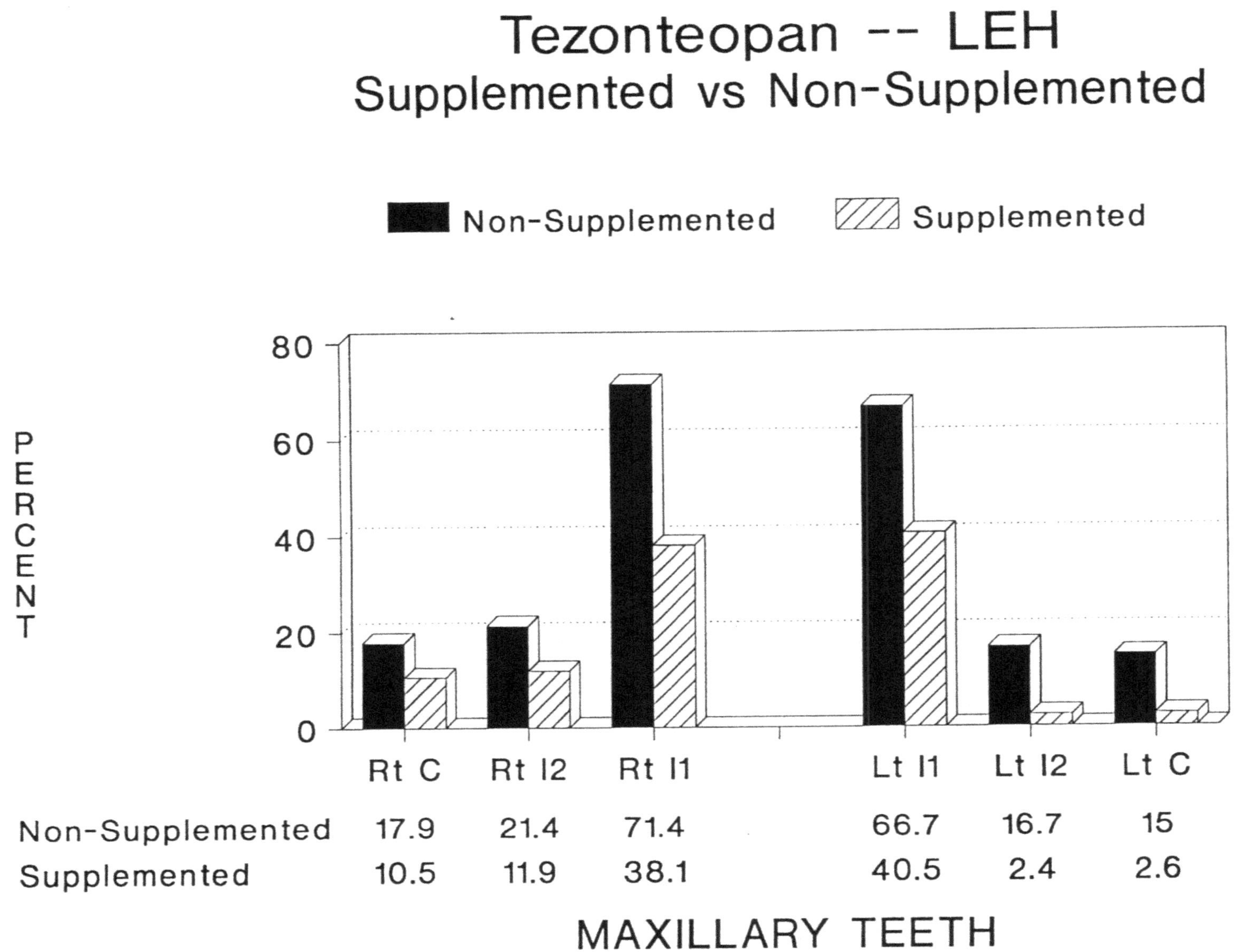

Tezonteopan -- LEH
Supplemented vs Non-Supplemented
Non-Supplemented
Supplemented
PERCENT
80
60
40
20
0
Rt C
Rt I2
Rt I1
Lt I1
Lt I2
Lt C
Non-Supplemented 17.9 21.4 71.4 66.7 16.7 15
Supplemented 10.5 11.9 38.1 40.5 2.4 2.6
MAXILLARY TEETH

Figure 8.8a

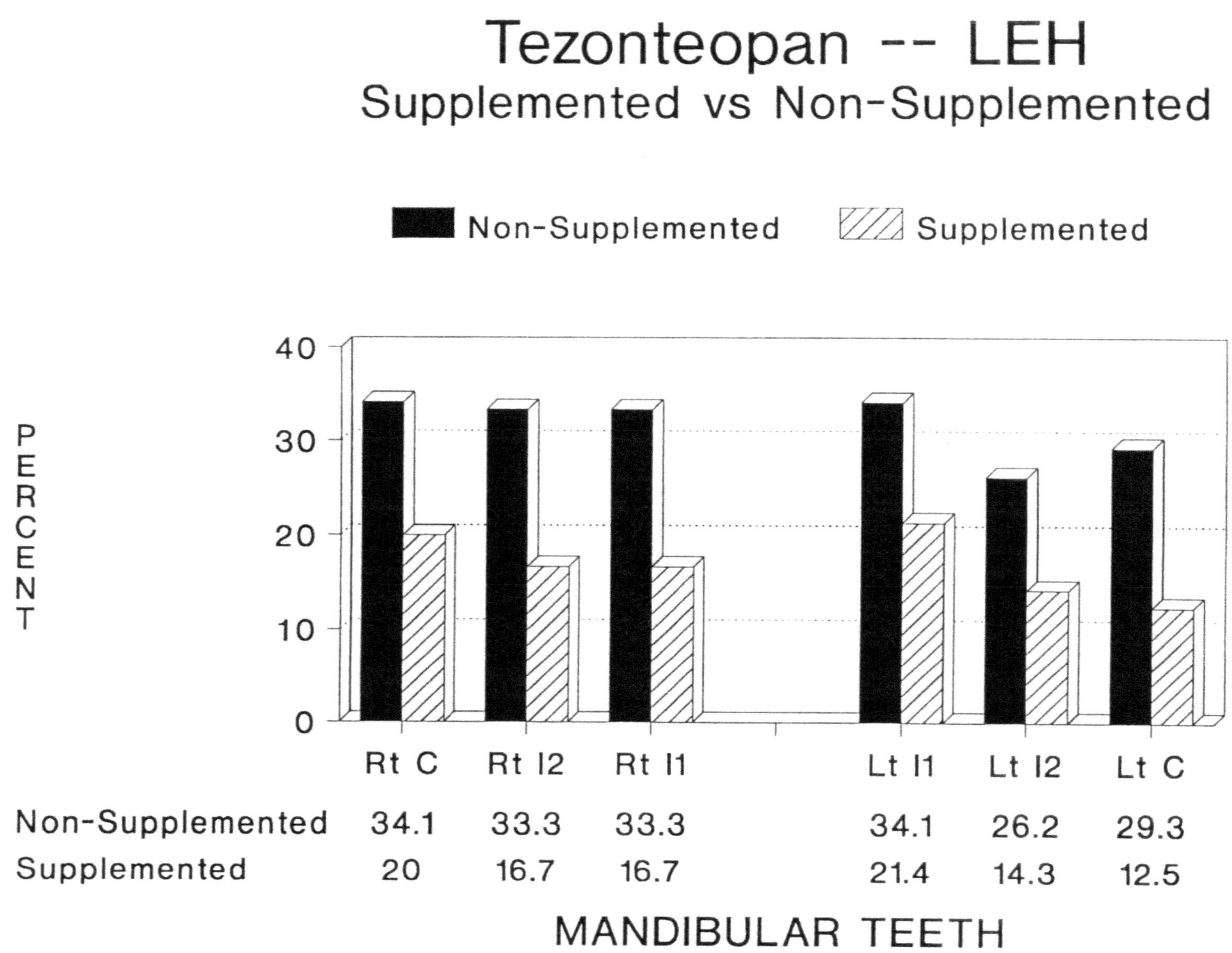
Tezonteopan -- LEH
Supplemented vs Non-Supplemented
Non-Supplemented
Supplemented
PERCENT
40
30
20
10
0
Rt C
Rt I2
Rt I1
Lt I1
Lt I2
Lt C
Non-Supplemented 34.1 33.3 33.3 34.1 26.2 29.3
Supplemented 20 16.7 16.7 21.4 14.3 12.5
MANDIBULAR TEETH

Figure 8.8b

TEZONTEOPAN -- LEH, MAX I1
Supplemented vs Non-Supplemented

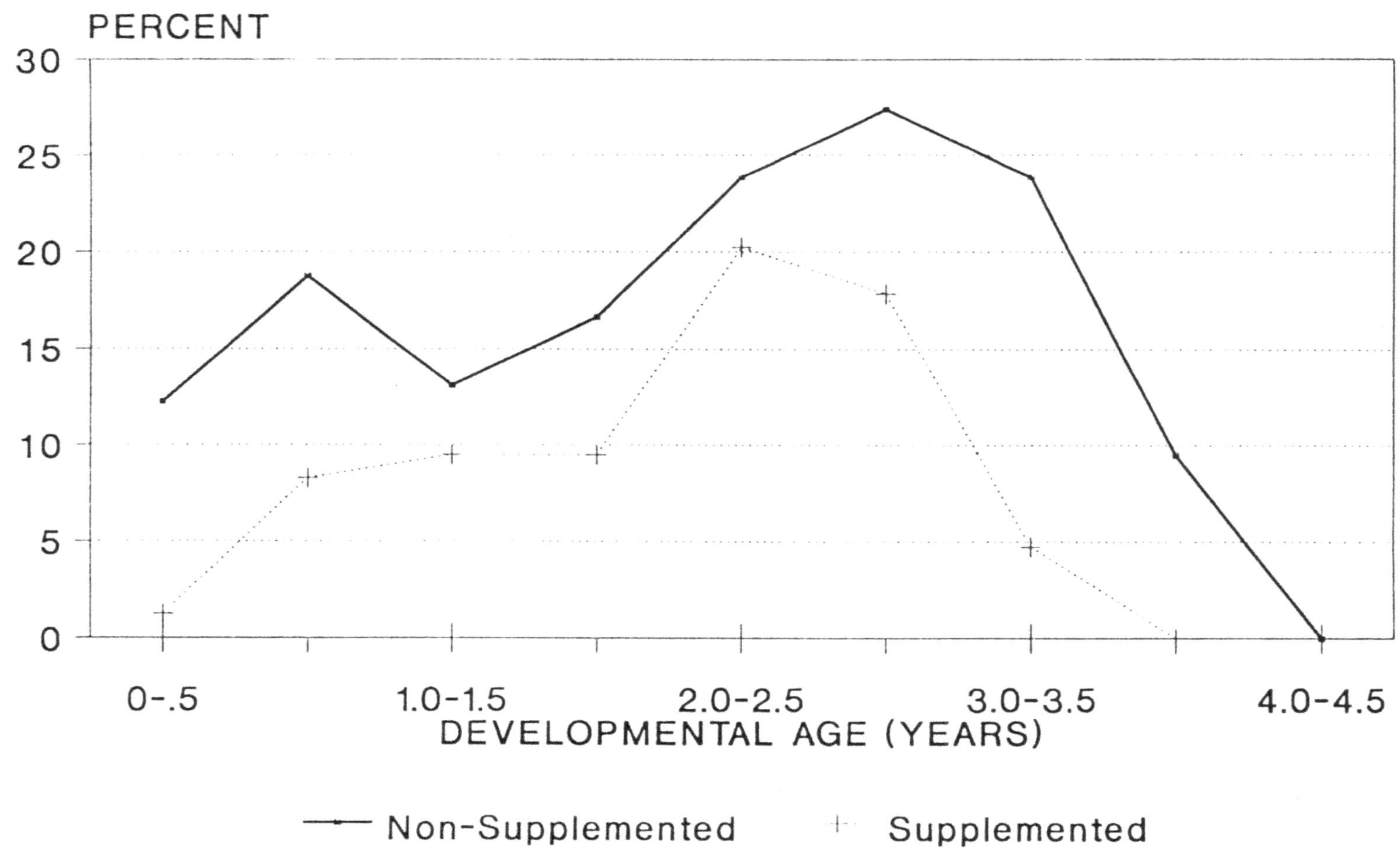

Note: Mean of Left and Right

Figure 8.9

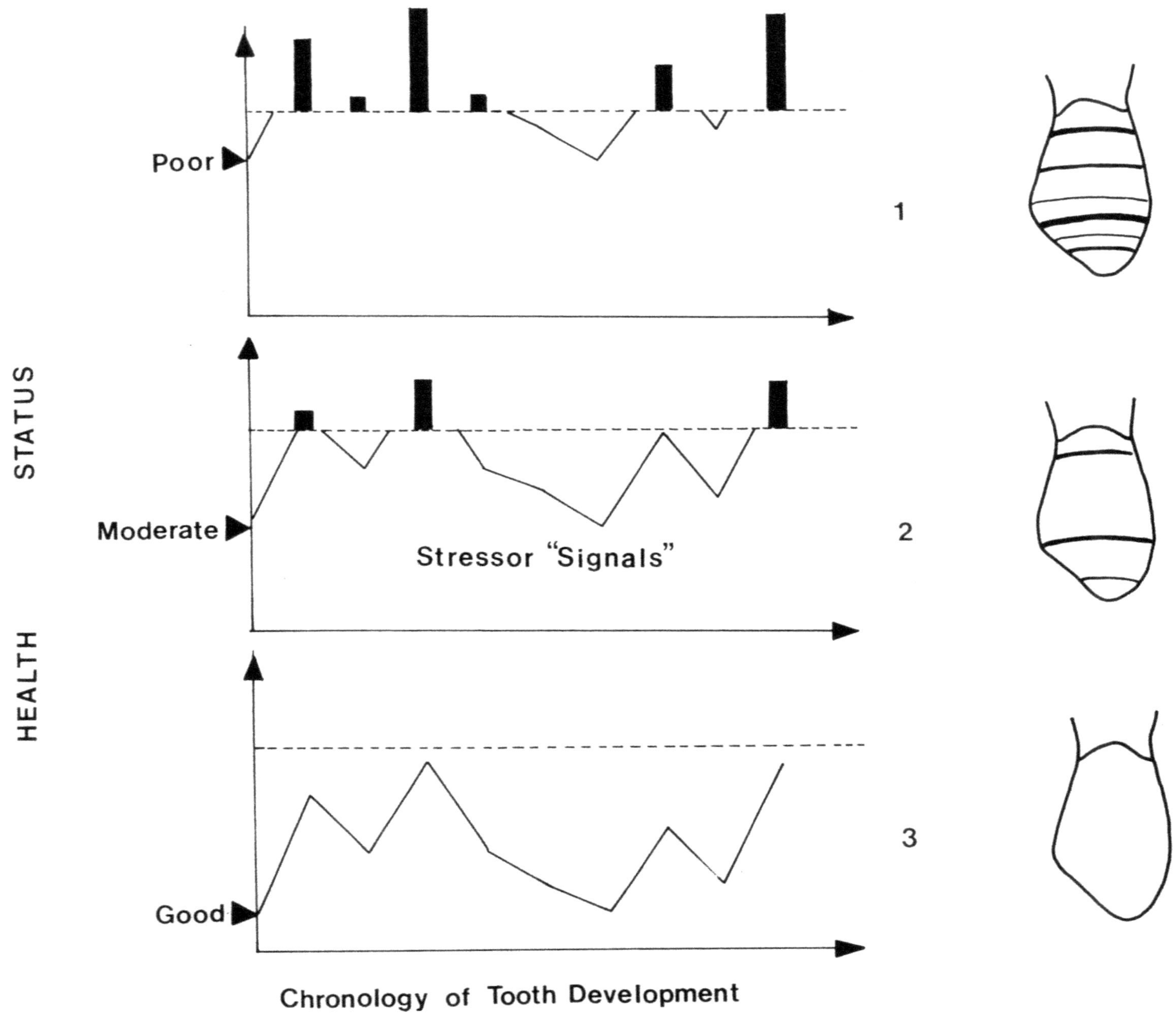

Figure 8.10 Hypothetical model showing different patterns of LEH produced by identical stressor events in individuals/populations of varying health status. The dotted line represents the threshold point a stressor signal must cross in order to produce a lesion.

9. ANAEMIA IN ROMAN BRITAIN: POUNDBURY CAMP.

Patty Stuart-Macadam

Introduction

Anaemia, or reduction of haemoglobin/ haematocrit in blood below normal levels, is a contemporary problem of world-wide significance. What was the story in the past? Fortunately anaemia is one condition that can leave its mark on bone. As a result, analyses of skeletal remains can provide information on its occurrence in the past. This, in turn, provides clues to the culture, environment, ecology and health status of past populations.

Anaemia can be recognized in skeletal material by the presence of a palaeopathology called porotic hyperostosis (Stuart-Macadam, 1982, 1987a). Porotic hyperostosis (after Angel, 1966) is a term that reflects the two-fold effect that the hyperplasia (increase in size) of bone marrow associated with anaemia has on bone:

1) a thinning and/or destruction of the outer table of compact bone in the form of porotic lesions

2) an increase in the middle layer of bone, or diploe (hyperostosis).

The lesions commonly affect the skull vault on the frontal, parietal and occipital bones (Figure 9.1), as well as the roof of the orbit (also known as cribra orbitalia) (Figure 9.2). The lesions have a symmetrical distribution and can vary in severity of expression from small porosities to large interconnected trabeculae (Figures 9.3a and 9.3b). It is well known that porotic hyperostosis occurs more commonly in juveniles, but this is probably a reflection of bone and marrow physiology rather than a real difference in incidence of anaemia between juveniles and adults (Stuart-Macadam, 1985). The lesions usually occur with about the same frequency in males and females (Stuart-Macadam, 1985), but since adult lesions most probably reflect childhood episodes of anaemia, this again does not give a true picture of anaemia in past populations. For these reasons, care must be taken in the interpretation of lesions of porotic hyperostosis seen in skeletal material. Even with these reservations, however, porotic hyperostosis is a useful indicator of health that can be used for cross cultural and temporal comparisons.

The present paper presents a comprehensive study of porotic hyperostosis in one skeletal collection from a Romano-British site, Poundbury Camp. It includes detailed descriptions of the lesions, as well as an examination of the factors associated with this particular population that may have contributed to the development of anaemia. In this way a picture is developed of anaemia in the past and of its affects on one Romano-British population.

Material and Techniques

Poundbury Camp is a 3rd-5th century Romano-British site situated in Dorset, S.W. England. Associated with the site was a large cemetery which occupied between two and four acres before its eventual abandonment in the fifth century. Excavations were carried out over a period of twelve years, from 1966 to 1978, under the direction of Chris Green. This resulted in the recovery of over 1000 human skeletons. Archaeological evidence indicates that the cemetery was used mainly during the fourth century, at which time it apparently served the adjacent Roman town of Durnovaria, present day Dorchester (Figure 9.4). The majority of burials were oriented along an east-west axis, with the head facing west (Figure 9.5). Generally there were no grave goods, but there were a number of special graves with stone, lead-lined or wooden coffins. It has been suggested that Poundbury Camp is one of the earliest Christian cemeteries to have been identified in Britain and one which indicates a large and well-organized community within a Roman town (Green, nd).

The Poundbury Camp skeletal collection is presently housed in the British Museum (Natural History). Between 1980 and 1982 the author collected data from the collection, including age, sex, macroscopic and radiographic features of porotic hyperostosis, metopism (retention of the medio-frontal suture) and dental enamel hypoplasia. Age at death of adults was estimated with the use of conventional morphological criteria (McKern and Stewart, 1957; Gilbert and McKern, 1971), fusion of the basilar suture, and dental attrition (Miles, 1963). The age of immature individuals was based on epiphyseal union and dental eruption sequence (Brothwell, 1972). Age estimates were made to within two-year intervals until the age of sixteen and thereafter within ten year intervals. Since the main focus of the study was porotic hyperostosis, only crania which were considered to be sufficiently complete for analyses were included. This made a total of 752 individuals. The objective of the study was to document thoroughly the occurrence of porotic hyperostosis, to try to determine its aetiology, and to identify contributing factors specific to Poundbury Camp. It was an ideal skeletal collection for this purpose because of its large sample size and the abundance of juvenile and infant remains.

The intra-observer error was calculated for both metrical and non-metrical data after repeating observations on successive days. The Students t-test was used as the statistic for comparing the means of two samples from the same population. Comparison of the proportions of two different groups was tested using X^2 that was corrected for continuity using Yates correction. In cases where the total sample size, N, was less than 20, or where N lay between 20 and 40 and the smallest expected number was less than 5, Fisher's exact probability was calculated.

Results

Of the 752 individuals examined, 38.8% (292) were adult females, 3.3% (254) were adult males, and 27.4% (206) were juvenile. Figure 9.6 illustrates the age distribution for the total collection.

The total number of skulls with porotic hyperostosis was 230 (30.6%). Of the 230 affected skulls, 75% (173) had cribra orbitalia only, 3% (7) had vault lesions only, and 22% (50) had both cribra orbitalia and vault lesions. Twenty seven point seven per cent(81) of females, 24.4% (62) of males and 36.4% (75) of juveniles had orbital lesions. Two point four per cent (7) of females, 10% (25) of males, and 12.1% (25) of juveniles had vault lesions. The juveniles showed a significantly higher incidence of both cribra orbitalia (X^2 = 5.6, p>.01) and vault lesions (X^2 = 6.7, p>.01) than adults. Table 9.1 and Figures 9.7 and 9.8 show the breakdown in age of individuals with both types of skull lesions. There was no significant difference between the male and female incidence of cribra orbitalia, but significantly more males (exact multinomial probability >.001) had vault lesions. The distribution of orbital and vault lesion types for the total group is provided in Figures 9.9 and 9.10. There was no significant difference in distribution of orbital lesion severity between males and females, but males had significantly more type 2 vault lesions than females (X^2 = 9.8, p>.001). Both cribra orbitalia (exact multinomial probability <.001) and vault lesions had a significantly greater severity in the juvenile group (exact multinomial probability >.01).

In 90% of cases the cribra orbitalia was symmetrical in expression and there was no significant difference between the left and right orbit in severity of the lesion. The location of cribra orbitalia on the orbit and the type of lesion is illustrated in Figure 9.11. Eighty-eight percent of the cribra orbitalia was localised in the antero-intermediate and antero-lateral areas of the orbit. The most common location was the antero-intermediate sector (50.8%), followed by the antero-lateral (37.2%) and the antero-medial (8.7%). No lesions of any kind were observed on the medial or lateral walls or the floor of the orbit.

The most common location for vault lesions was the parietal area (48.2%), followed by the occipital (36.1%), and frontal (15.7%). In 61% of cases the vault lesions occurred on one area only, and in 39% of cases the lesions occurred in two or more areas. Eighty-eight percent (45 of 51) of vault lesions were associated with orbital lesions while only 27% (45 of 167) of orbital lesions were associated with vault lesions. Vault lesion types 1, 4 and 5 (Figures 9.3a and 9.3b) were always associated with orbital lesions.

Analyses of radiographs of those individuals who had porotic hyperostosis showed that there were bone changes consistent with those seen on radiographs of modern individuals with anaemia. Skull radiographs showed a six percent incidence of "hair-on-end" trabeculation, orbital roof thickening and slight to moderate increases in the diploe. Radiographs of long bones of juveniles with porotic hyperostosis showed some thinning of the cortex and widening of the medullary cavity, which again was consistent with bone changes seen in modern cases of anaemia.

Discussion

The presence of porotic hyperostosis in the human remains from Poundbury Camp indicates that anaemia was a condition experienced by many individuals during the course of their lifetime. Was the anaemia environmental or genetic? What factors may have contributed to the development of this anaemia? A consideration of factors that can produce anaemia with reference to the geography, demography, culture and environment of this particular Romano-British group should provide answers to these questions.

a) Diet

Information on diet and nutrition in Roman Britain is minimal in the literature. According to Birley (1964) the Romanized Britons adopted Roman practice in taking their meals; this would have meant a breakfast of bread and fruit, a light lunch of fish, eggs, vegetables and wine or beer and a main meal of three parts. The first part included eggs, shellfish and salads with wine; the second was meat, usually beef or mutton; and the last included puddings, pastries, cakes, fresh and dried fruit and wines. Wheat, barley oats and rye were used to make both leavened and unleavened bread. There was a good variety of fruit and vegetables, including some items that were introduced by the Romans.

From this description it appears that the average diet was quite adequate with regard to nutrients and minerals, including iron. However, most likely this Romanised diet was that of the well-off or upper class individuals. It has been said (Woodman, 1978) that the ordinary native Briton would have seen little change in diet after the Roman occupation; most people subsisted on flat bread made from coarse grain flour, and bean pottage or porridge with the occasional addition of meat. This assessment of diet is supported by examination of Poundbury Camp dentition (Lonsdale, 1981). On the basis of the caries rate and condition of teeth Lonsdale concluded that Poundbury Camp individuals had consumed a coarse diet, and one which included either sweet substances such as honey or sticky starchy food such as porridge. Other evidence from the Poundbury Camp site shows that wheat, barley and oats were included in the diet (Green, nd). This sort of diet could lead to a population much more susceptible to deficiencies of various vitamins and minerals. However, the absence of obvious signs of deficiency diseases and the faunal evidence indicating utilisation of a wide variety of animals including ox, sheep, goat, pig, red and roe deer (Green, nd) lead to the conclusion that the diet of Poundbury Camp individuals was probably adequate in terms of nutrients and essential minerals, including iron.

b) Parasites

In Britain worms have been implicated in cause of death and certainly associated with intestinal upsets since the 1600's (Fountain, 1961). The types of helminths noted in early registrar's reports from London include tapeworms, roundworms, threadworms, guinea worms and ydatids. In 1840 it was noted that "other parasitic animals and indeed those above named, though frequent complication of infantile diseases, rarely prove fatal".

Parasite eggs have been found in material associated with Poundbury Camp; Jones (pers. comm.) located trichurid or whipworm eggs in samples from the site. Whipworms bore into the wall of the large intestine so that the anterior extremity is in the mucosa (Davey, 1961). In humans infection can take place when parasite eggs containing infective larvae are swallowed. This can occur for several reasons, including pollution of drinking water or food, especially green vegetables or contamination of hands by infected soil and transfer of eggs to the mouth. Although effects are not serious, diarrhoea and mild gastric disturbances can occur (Davey, 1961). Obviously an affected individual would be prone to gastric upsets.

It is not known whether the Poundbury Camp population was exposed to infestation from any other types of helminths. It is possible that any of those mentioned in the early London Registrar's reports were also present during the fourth century, and many of them cause intestinal upsets. Roundworms, which live in the small intestine, can cause intestinal obstruction, and tapeworms can produce a type of anaemia associated with gastric disturbance.

c) Malaria

For centuries malaria was endemic in certain areas in England, particularly the Fen country, marshes of the Thames estuary and S.E. Kent, the low lying country around Bridgewater (Somerset) and the Ribble District of Lancashire (MacArthur, 1951). However, serious epidemics did not occur because even summer temperatures were rarely high for long enough to permit the parasite to undergo sexual reproduction in the mosquito to produce sporozoites (Clegg, 1973). The species that were prevalent in Britain were *Plasmodium vivax* and *malariae*, whose sexual cycles are completed at lower temperatures than those of other species. Malaria resulting from these *plasmodia* is not as serious or frequently fatal as that due to *Plasmodium falciparum*. It is though that because of this and the low proportion of the total population that was exposed to malaria that there was not much population selection by malaria (Livingstone, 1967).

The possibility that malaria occurred in Roman times at Poundbury Camp was considered, as this area of Dorset is not too distant from Bridgewater, where malaria has been documented. However, according to Henschen (1961), malaria did not spread throughout Europe until after the return of the crusaders in the twelfth century. Evidence for this assertion is not provided, but if it is correct, then malaria would not have been a factor contributing to the health status of these earlier Romano-British people.

d) Lead

The occurrence of chronic lead poisoning is a very real possibility in Roman Britain. The Romans had developed an extensive lead technology that utilized lead in a variety of ways such as pipe lining, in cooking utensils, storage vessels for beverages, coins, toys, household repairs, cosmetics and even coffins (Gilfillan, 1965). Lead was also incorporated into wine in as many as fourteen different ways, including as a preservative and to improve flavour. In classic experiments in the 1800's, Hoffman (c.f. Steinbock, 1976) boiled various wines with lead according to Roman instructions and found that extremely high lead levels resulted.

The full extent of Roman influence on the indigenous British population in terms of lead use is not known. Poundbury Camp cemetery, however, was adjacent to and served the large Roman town of Durnovaria. Lead was known to have been mined by the Romans in occupied Britain and in fact, one of the mines in the central Mendips was north of Poundbury Camp. That lead was in use during this time is substantiated by the presence of lead lined coffins in the cemetery and by evidence for lead working on the site (Green, nd). Unfortunately, the full extent of lead use is difficult to assess as no grave goods have been recovered and the town itself has not been excavated.

Waldron *et al* (1976, 1979) have carried out extensive studies on the Poundbury Camp skeletal material to determine lead levels. A number of the samples contained 100 ppm or more of lead; some even have concentrations in excess of 200 ppm. Compared with results obtained from other archaeological specimens, these appear to be extremely high levels. The mean concentration of lead found in Poundbury Camp bones is 112.7 ppm, which is significantly higher ($p<.001$) than those from Henley Wood, a Romano-British site in Somerset (Waldron *et al*, 1976). Taking levels in modern bone into consideration, Waldron *et al* (1976) found that the Poundbury Camp sample mean is greater by a factor of more than two. The bones showed a weak positive correlation between lead content and age, which would be consistent with the observations of lead metabolism in modern clinical studies (Molleson, pers. comm.).

The soil environment was considered as a possible influence on bone lead levels. Studies have indicated that the transport of lead in all but acidic soils is small (Grandjean and Holma, 1973; Jurinak and Santillan-Medrano, 1974). The subsoil at Poundbury Camp consists of Upper Chalk and the grave fills are chalk rubble and occasional lenses of grey-brown soil (Waldron *et al*, 1979). These are alkaline soils that would not encourage lead transfer. The correlation between concentration of soil lead and bone lead is of a very low order, supporting an absence of lead transfer.

Waldron *et al* (1976) felt that the lead levels in the children at Poundbury Camp are so unusually high that lead poisoning, either acute or chronic, may have played a part in death. Certainly the Poundbury Camp population appears to have been exposed to levels of environmental lead that were much greater than in other areas at similar time periods and than those found today. Since lead poisoning can be associated with anaemia (Waldron, 1966) the possibility of a correlation between the presence of porotic hyperostosis and high lead levels was investigated. The results did not indicate a correlation between the two variables. This was not considered to invalidate the presence of either anaemia or high lead levels since there are so many unknown factors involved in the pattern of skeletal lead retention and the development of bone lesions due to anaemia.

e) Evidence for genetic anaemia

According to Livingstone (1967) abnormal haemoglobins[1] are apparently quite rare in all populations of northern Europe. Large surveys in England have found only trace frequencies of abnormal haemoglobins and thalassaemia[2] (Livingstone, 1967). While it is likely that haemoglobinopathies[3] were present in fourth century Romano-Britons, it is unlikely that the incidence would have been appreciably different to today. Not a single case of genetic anaemia has been recognised in any British skeletal population examined to date. Although the Romans came from an area where at present there is a high frequency of occurrence of thalassaemia, it is doubtful whether their numbers in Britain would have significantly altered gene frequencies in the indigenous population.

What type of anaemia was affecting Poundbury Camp individuals? Modern studies indicate that the incidence of abnormal haemoglobins is very low in northern Europe and most likely this was the case in the past. Radiographic analyses of skulls of individuals from Poundbury Camp with porotic hyperostosis showed that there was only slight to moderate hyperplasia of the bone marrow in the skull and long bones. There was no evidence of any of the more severe skeletal changes diagnostic of the genetic anaemias such as extreme thickening of the skull vault, facial changes, or hand and foot changes in juveniles. The greatest number and the most severe lesions occur in age groups between six months and four years, although it should be remembered that this is also the time period when anaemia is most likely to result in bony change (Stuart-Macadam, 1985). Nevertheless, this data from the skeletal material and the modern distribution of abnormal haemoglobines supports a diagnosis of an acquired anaemia such as iron deficiency anaemia, where increases of marrow hyperplasia are only slightly greater than normal and younger children are very commonly affected.

Iron deficiency anaemia is a general descriptive term for anaemia that is acquired, that is associated with slight increases in marrow hyperplasia, and that is associated with lower than

---

1 abnormalities in the oxygen-carrying pigment of the red blood cells.

2 a group of genetically determined disorders characterized by deficient haemoglobin synthesis.

3 haemtatologic disorders due to alterations in the genetically determined molecular structure of haemoglobin.

normal levels of haemoglobin and haematocrit[4] in the blood. There are a number of factors that can cause this type of anaemia, including blood loss (haemorrhage, parasite infestation), deficiency in the diet during periods of accelerated demands, inadequate absorption of iron (because of diarrhoea), or nutritional deficiencies (Robinson, 1972). What factors may have been responsible for the anaemia at Poundbury Camp? The absence of any obvious signs of nutritional deficiencies as well as information on the flora and faunal remains from the Poundbury Camp site do not support a dietary causation for the iron deficiency anaemia. It is becoming more apparent that, except in extreme cases of malnutrition, dietary explanations for iron deficiency anaemia are not satisfactory (Stuart-Macadam and Kent, in press). According to Passmore and Eastwood (1986:459) "it is unusual for anaemia to arise in an otherwise healthy person solely as a direct result of poor diet".

Parasitism can affect iron metabolism either by direct blood loss or by causing symptoms such as diarrhoea which inhibit iron absorption. There is evidence for the presence of parasites at Poundbury Camp and these certainly could have been a contributing factor in the occurrence of anaemia. Acute and chronic diseases are also associated with a type of iron deficiency anaemia, as the organism attempts to withhold iron from the invading organism. This type of anaemia could have affected any member of the population, but younger juveniles would have been particularly affected, since their system is still developing and susceptible to infectious and viral disorders. Lead poisoning can lead to anaemia, as well as a number of other problems. The mechanism is not completely understood, but it appears that lead interferes with the synthesis of haem, which is an essential component of haemoglobin (Wintrobe, 1974). Adults are usually not severely affected, but children can be. To what extent lead anaemia may have been a factor at Poundbury Camp is unknown, but Waldron's research suggests that it did occur.

Evidence from the bones, as well as environmental and cultural data, indicates that the type of anaemia that was affecting the Poundbury Camp was an iron deficiency anaemia. What does this tell us about this Romano-British population? How does it compare with other populations which differ chronologically and geographically? The data reveals that 26% of Poundbury Camp adults and 36% of juveniles show signs of anaemia. It is actually quite difficult to compare data from the literature on porotic hyperostosis with the Poundbury Camp data. Firstly, in other studies the type of anaemia which affected the population, whether acquired or genetic, is not always fully explored. Secondly, the age distribution of a population can greatly affect the apparent incidence of porotic hyperostosis. For example, iron deficiency anaemia is much more common in the six month to two year age period in modern times and probably showed a similar pattern in the past. This younger age period is also much more likely to exhibit bony lesions associated with anaemia simply because of bone and marrow physiology (Stuart-Macadam, 1985). As a result of these two factors, a population with a proportionately larger number of younger juveniles will have a higher incidence of porotic hyperostosis than a population with fewer younger juveniles. Thirdly, there is the problem of determining the difference between active and healed lesions which has never been conclusively addressed. Some studies attempt to differentiate between these, others do not. However, it is very difficult to compare data when this particular aspect of porotic hyperostosis has not been standardized. A consideration of bone and marrow physiology leads to the conclusion that lesions of porotic hyperostosis seen in adults represents childhood episodes of anaemia (Stuart-Macadam, 1985). As a result, adult porotic hyperostosis is not an accurate indicator of adult anaemia. Therefore, to learn about the pattern and aetiology of anaemia in the past it is more biologically meaningful to focus on the juveniles. With all these reservations in mind, what does the pattern of porotic hyperostosis in juveniles from Poundbury Camp tell us?

Porotic hyperostosis occurs in 36% of the juveniles from Poundbury Camp and is most common and severe in the six month to four year age categories. It is important to be aware that this figure underestimates the true number of juveniles who would have had iron deficiency anaemia. Modern clinical studies have shown that at most, 50-75% of patients with various sorts of anaemias exhibit bony change which can be detected on radiographs (Stuart-Macadam, 1982). In some cases even individuals with severe anaemia never develop bone alterations. This suggests that more than one-third of all juveniles at Poundbury Camp had anaemia at some time in their lives. Data on 1,533 juveniles taken from a number of studies indicates that the average incidence of porotic hyperostosis in the past was 36% and the range varied from 26.5% to 83% (Stuart-Macadam, 1985). It should be noted that some of the sample sizes were very small, and the age distribution would be variable, which would affect the incidence. However, these figures show that Poundbury Camp juveniles had no more and no less anaemia than an average population from various parts of the world. Data from another Romano-British cemetery, Cirencester, shows that 35.1% of the juveniles had porotic hyperostosis (Wells, 1982). So, at

[4] the volume percentage of red blood cells in whole blood.

least in terms of producing anaemia, conditions at both Romano-British sites must have been similar.

The Poundbury Camp data showed that more severe lesions (linked foramina and outgrowth from the normal contour of the outer table) were seen only in individuals under the age of fourteen years and the greatest frequency of severe lesions occurred in the six month to two year age group. There is no published information on severity of lesions in juveniles from other Romano-British populations, so it is not possible to make comparisons. Although the incidence of cribra orbitalia is similar between males and females at Poundbury Camp, there is a greater frequency of vault lesions of porotic hyperostosis in males. If vault lesions represent a more severe episode of anaemia, then this could indicate that boys are more susceptible to severe anaemia than girls. There is no comparative data of this sort from other studies, so it is not possible to make comparisons with other sites.

Conclusions

Analysis of porotic hyperostosis in individuals from Poundbury Camp provides information on the occurrence and aetiology of anaemia in this Romano-British population. The anaemia was most likely an acquired iron deficiency type, and the causative factors may have included parasitism, infectious diseases, and lead poisoning. It is not possible to make statements about anaemia acquired in adulthood, but at least one-third of the juveniles had anaemia during their lifetime. Younger juveniles seem to be more severely affected than older juveniles and there is a possibility that young boys may be more severely affected than young girls. For a number of reasons there are problems with comparing data from Poundbury Camp with data from other sites. However, it appears that the incidence of anaemia at Poundbury Camp was about average for prehistoric populations in general and similar to that seen at Cirencester, a contemporaneous Romano-British population.

References

Angel, J.L. (1966). Porotic hyperostosis, anemias, malarias and marshes in prehistoric Eastern Mediterranean. *Science 153*:760-63.

Birley, A. (1964). *Life in Roman Britain.* London: B.T. Batsford Ltd.

Brothwell, D. (1972). *Digging up Bones.* London: British Museum (Natural History).

Clegg, A.G. (1973). *Man Against Disease.* London: Heinemann Educational Books.

Davey, T.H. (1961). *A Guide to Human Parisitology.* London: J.K. Lewis and Co. Ltd.

Fountain, A.E. (1961). Records of mortality from helminth diseases in England and Wales up to 1880. *J. Helminthol. R.T. Leiper supplement*:77-84.

Gilbert, B.M. and McKern, T. (1971). A method for ageing the female *os pubis*. *Am. J. Phys. Anthropol. 38*:31-38.

Gilfillan, S.C. (1965). Lead poisoning and the fall of Rome. *J. Occup. Med. 7*:53-60.

Grandjean, P. and Holma, P. (1973). A history of lead retention in the Danish population. *Environ. Physiol. Biochem. 3*:268-73.

Green, C. (nd). *A summary of excavations at Poundbury, Dorchester, Dorset carried out between 1966 and 1976.* Unpublished.

Henschen, P. (1961). Cribra cranii - a skull condition said to be of racial or geographical nature. *Path. Microbiol. 24*:724-29.

Jurinak, J.J. and Santillan-Medrano, J. (1974). The chemistry and transport of lead and calcium in soil. *Researcher Report, 18*: Utah Agricultural Experiment Station, Logan, Utah State University.

Livingstone, F. (1967). *Abnormal Haemoglobins in Human Populations.* Illinois: Aldine Pub. Co.

Lonsdale, D. (1981). *A Study of the Dentition of a Romano-British Population with Particular Reference to Agenesis of Third Molars and Attrition.* Unpublished M.Sc. dissertation, University of Oxford.

MacArthur, W. (1951). A brief story of English Malaria. *Brit. Med. Bull. 8*:76-9.

McKern, T.W. and Stewart, T.D. (1957). Skeletal age changes in young American males, analyzed from the standpoint of identification. *Technical Report EP-45.* Natick, MA: Quartermaster Research and Development Center, Environmental Protection Research Division.

Miles, A.E.W. (1963). The dentition in the assessment of individual age in skeletal material. In *Dental Anthropology* ed. D.R. Brothwell. pp.191-209. Oxford: Pergamon Press.

Passmore, R. and Eastwood, M.A. (1986). *Davidson and Passmore Human Nutrition and Dietetics.* New York: Churchill Livingstone.

Robinson, C.H. (1972). *Normal and Therapeutic Nutrition.* New York: Macmillan.

Steinbock, R.T. (1976). *Paleopathological Diagnosis and Interpretation.* Springfield, Illinois: Charles C. Thomas.

Stuart-Macadam, P. (1982). *A Correlative Study of a Palaeopathology of the Skull.* Unpublished Ph.D. thesis, Department of Anthropology, University of Cambridge.

Stuart-Macadam, P. (1985). Porotic hyperostosis: representative of a childhood condition. *Am. J. Phys. Anthropol. 66*:391-98.

Stuart-Macadam, P. (1987a). New evidence to support the anemia theory. *Am. J. Phys. Anthropol. 74*:521-26.

Stuart-Macadam, P. (1987b). A radiographic study of porotic hyperostosis. *Am. J. Phys. Anthropol. 74*:511-20.

Stuart-Macadam, P. (1989). Porotic hyperostosis: relationship between orbital and vault lesions. *Am. J. Phys. Anthropol. 80*:187-95.

Stuart-Macadam, P. and Kent, S. (in press). *Diet, Demography and Disease: Changing Perspectives of Anemia.* Stanford: Stanford University Press.

Waldron, H.A. (1966). The anemia of lead poisoning: a review. *Br. J. Ind. Med. 23*:83-100.

Waldron, H.A., Khera, A., Walker, G., Wibberley, J.G. & Green, C. (1979). Lead concentrations in bone and soil. *J. Arch. Sci. 18*:221-98.

Waldron, H.A., Mackie, A. & Townshend, A. (1976). The lead content of some Romano-British bones. *Archaeometry 18*:221-27.

Wells, C. (1982). The human burials. In *Romano-British Cemeteries at Cirencester.* eds. A. McWhirr, L. Viner and C. Wells. pp.135-201. Cirencester: Cirencester Excavation Committee.

Wintrobe, M. (1974). *Clinical Hematology.* Philadelphia: Lea and Febiger.

Woodman, M. (1978). *Food and Cooking in Roman Britain.* Cirencester: Corinium Museum.

TABLE 9.1

INCIDENCE OF CRIBRA ORBITALIA AND VAULT LESIONS WITH AGE

Age (years)	Cribra Orbitalia	%	Vault Lesions	%
6 MONTHS-2 YEARS	15/35	43	9/35	25.7
2-4	13/15	87	6/15	40
4-6	10/21	47.6	5/21	23.8
6-8	5/10	50	0/10	0
8-10	11/14	79	2/14	14.2
10-12	6/13	46	0/13	0
12-14	6/8	75	0/8	0
14-17	5/12	41.6	0/12	0
17-25	40/101	40	10/98	10.2
25-35	43/174	24.7	12/175	6.8
35-45	41/168	24.4	5/165	3
45+	19/105	18	4/105	3.8

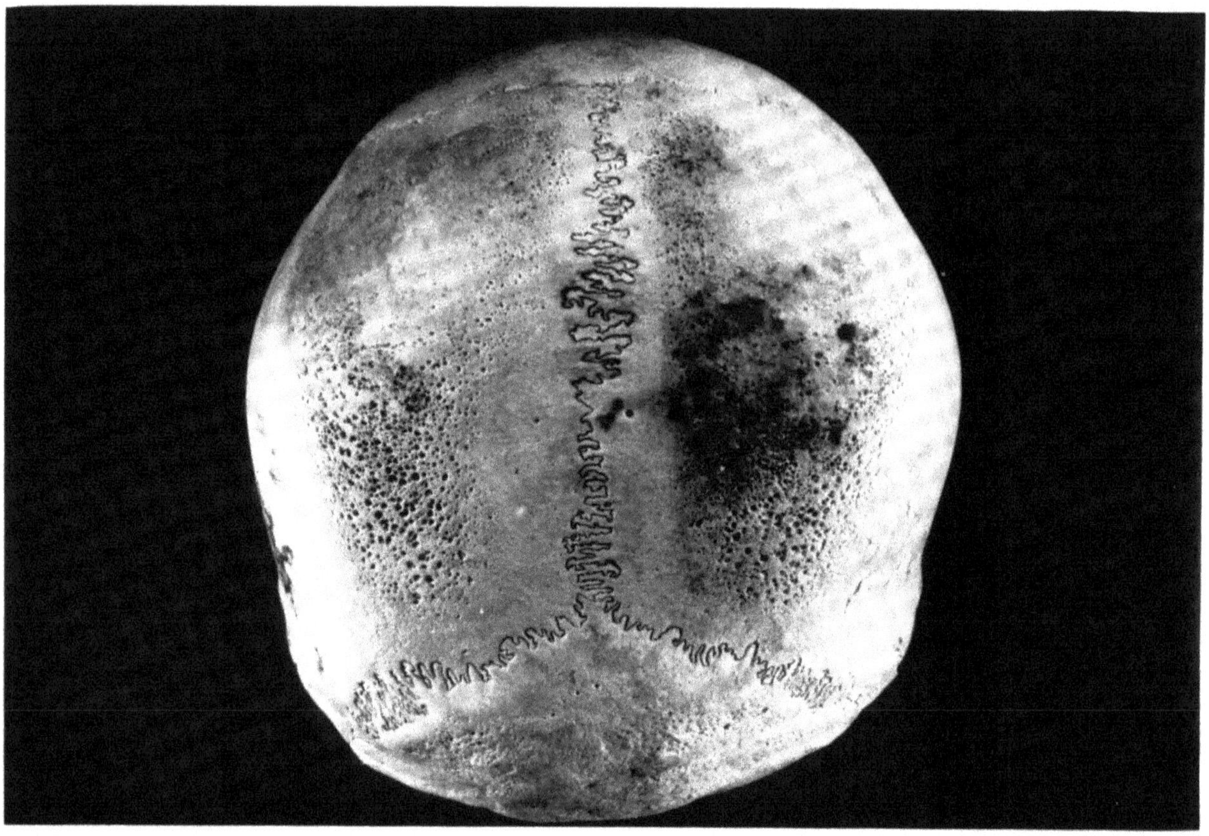

Figure 9.1 Vault lesions of porotic hyperostosis.

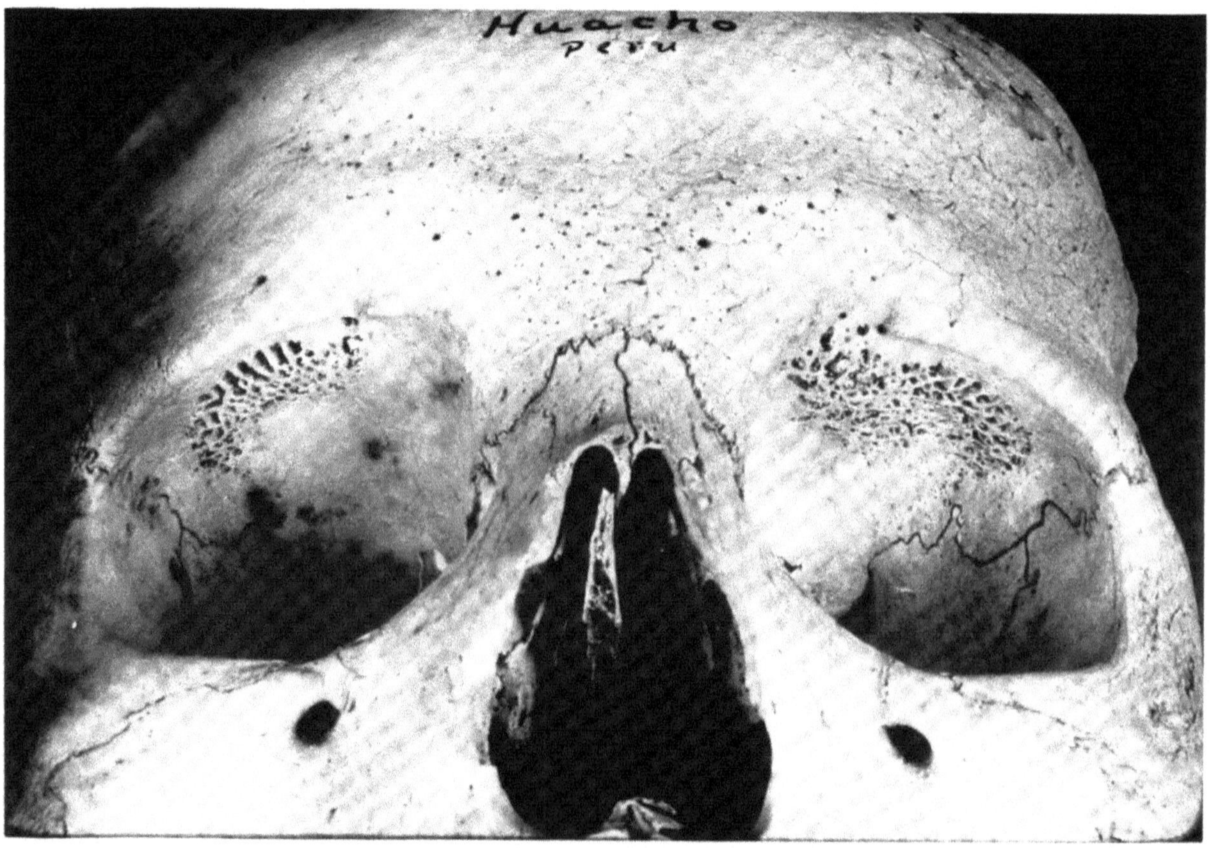

Figure 9.2 Orbital lesions (cribra orbitalia) of porotic hyperostosis (Peruvian origin).

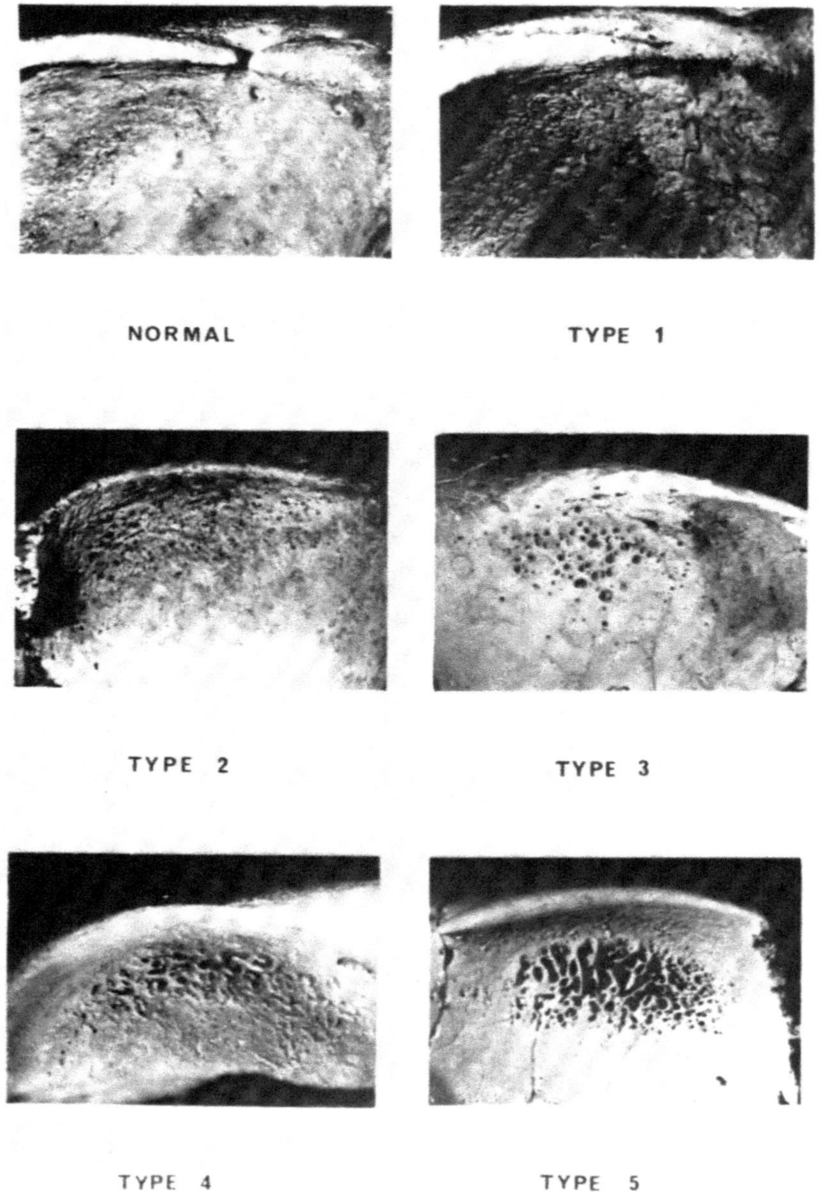

Figure 9.3a Illustrations of lesion types.

0 = normal bone surface
1 = capillary-like impressions on the bone
2 = scattered fine foramina
3 = large and small isolated foramina
4 = foramina have linked into a trabecular structure
5 = outgrowth in trabecular form from the outer table surface

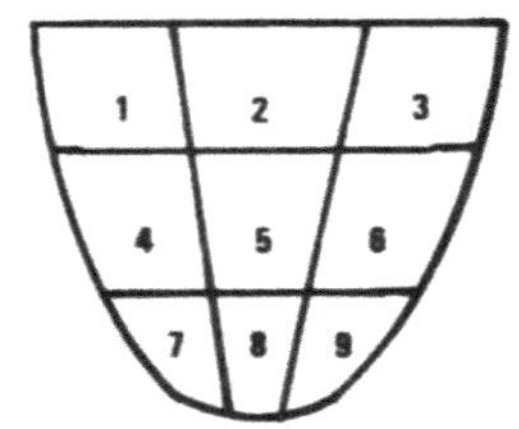

1 = Antero-lateral sector
2 = Antero-intermediate sector
3 = Antero-medial sector
4 = Middle-lateral sector
5 = Middle-intermediate sector
6 = Middle-medial sector
7 = Postero-lateral sector
8 = Postero-intermediate sector
9 = Postero-medial sector

Figure 9.3b Description of lesions and areas of orbit.

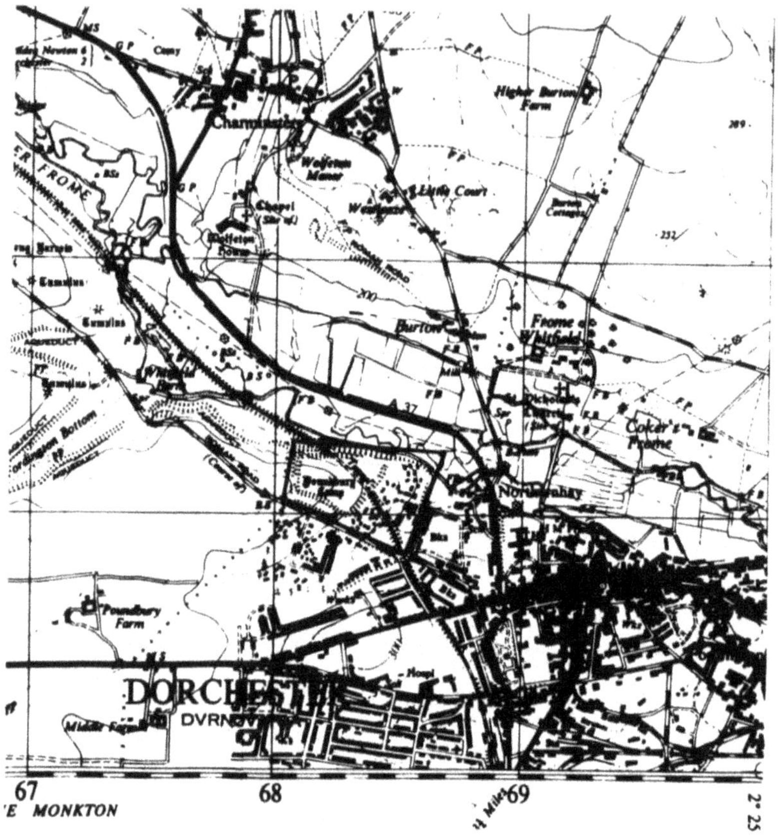

Figure 9.4 Map of Dorchester showing Poundbury Camp location.

Figure 9.5 Grave distribution at Poundbury Camp.

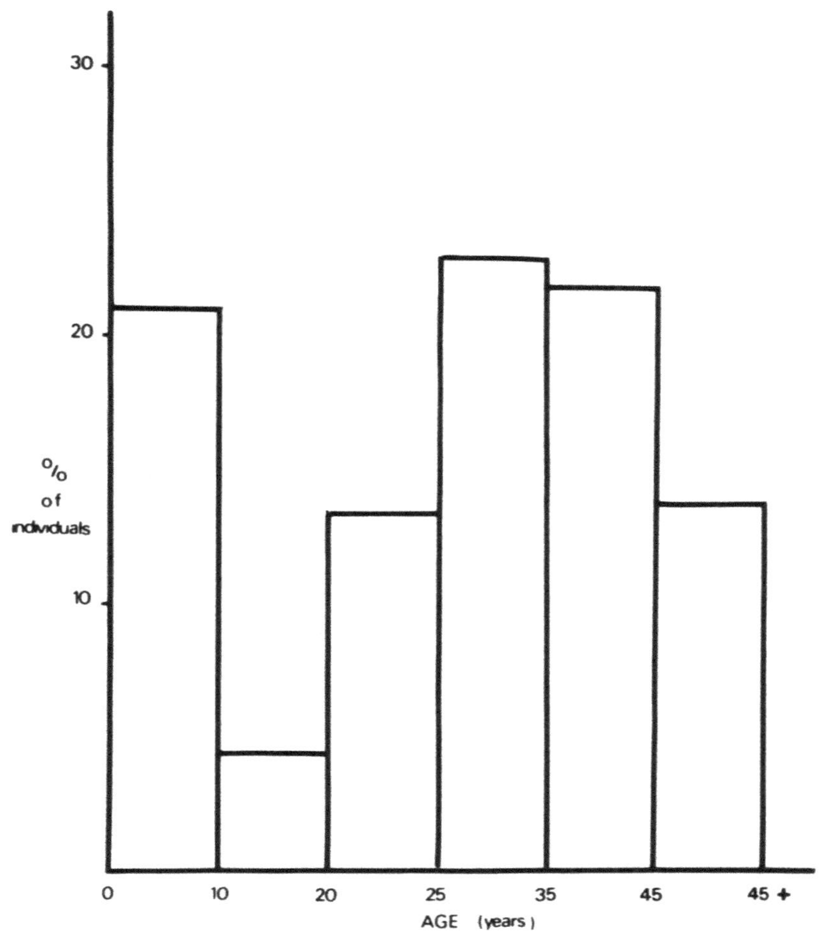

Figure 9.6 Age distribution for Poundbury Camp individuals.

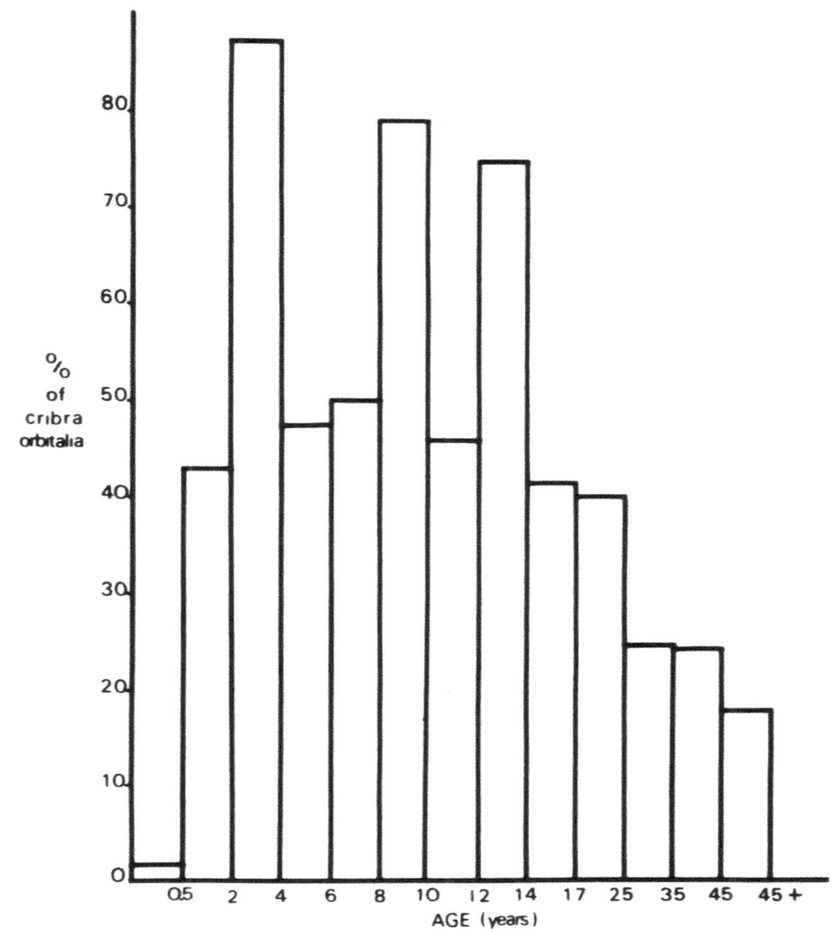

Figure 9.7 Distribution of cribra orbitalia with age.

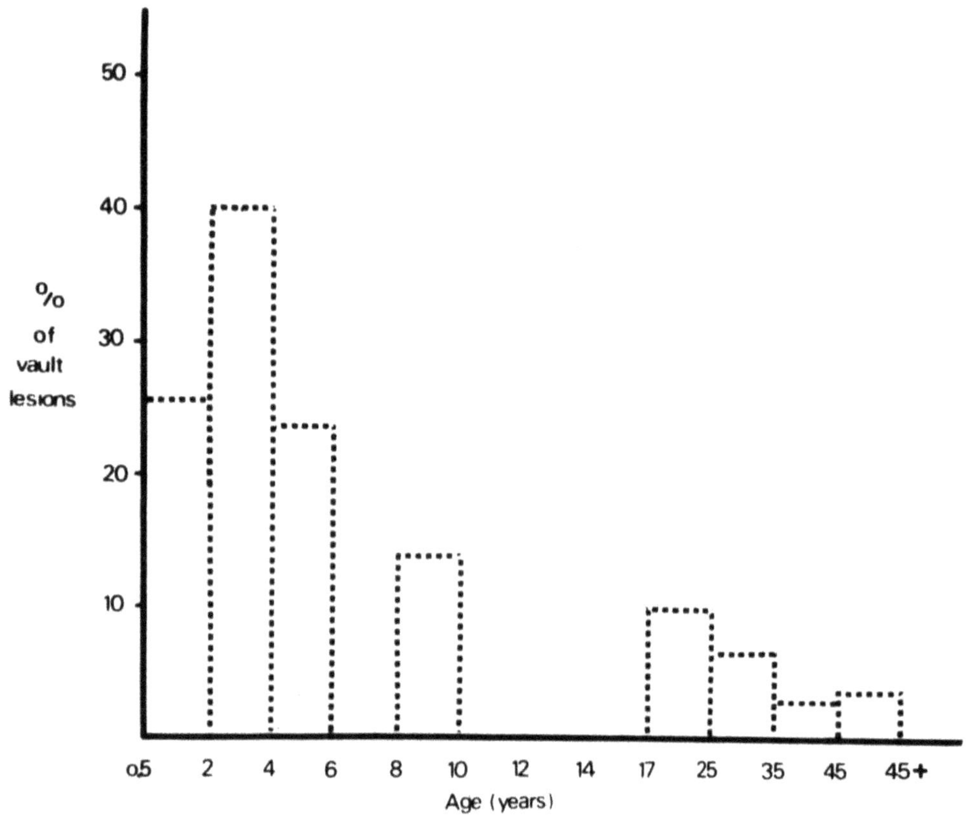

Figure 9.8 Distribution of vault lesions with age.

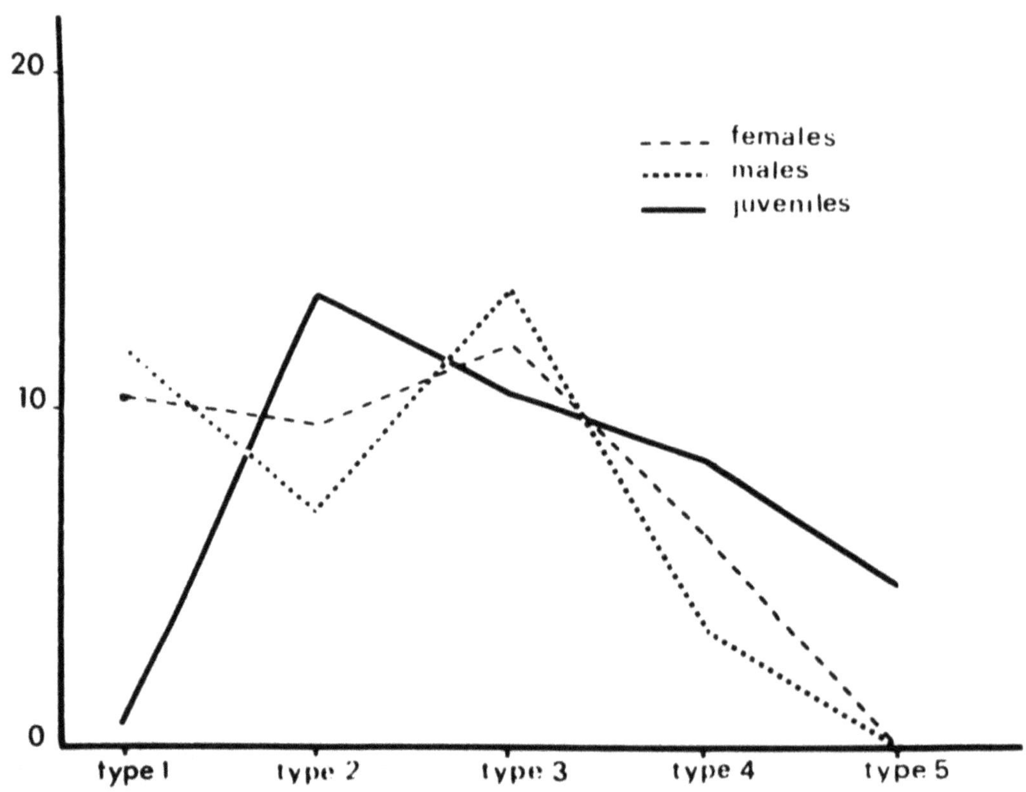

Figure 9.9 Orbital lesion type in Poundbury Camp individuals.

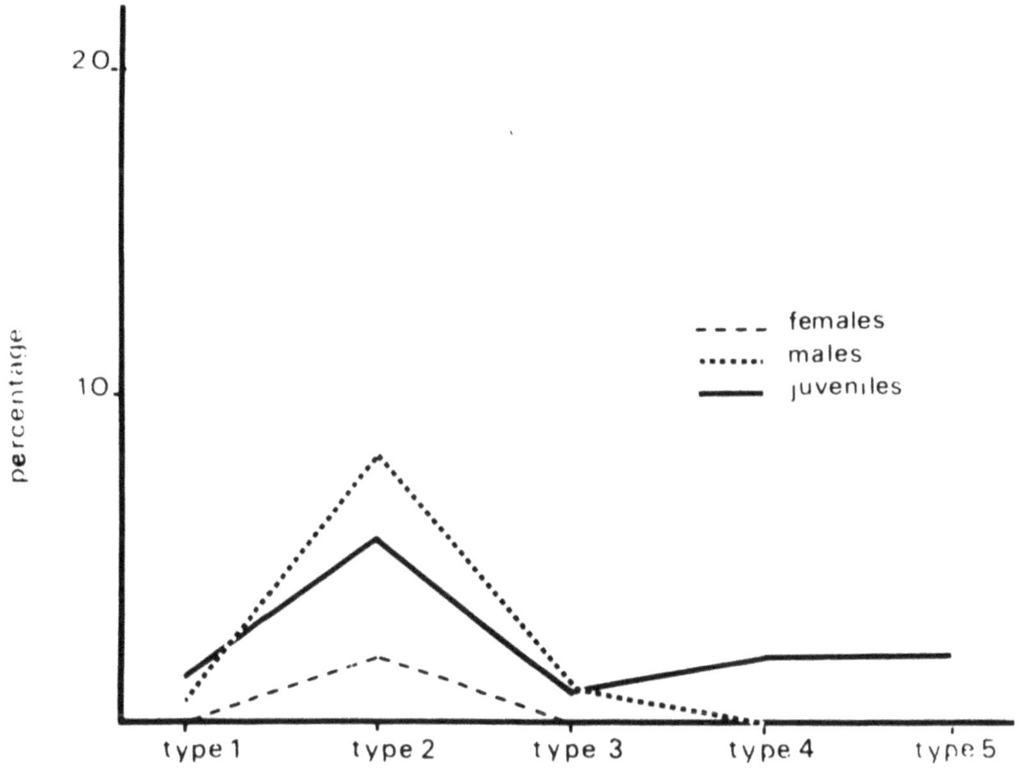

Figure 9.10 Vault lesion type in Poundbury camp individuals.

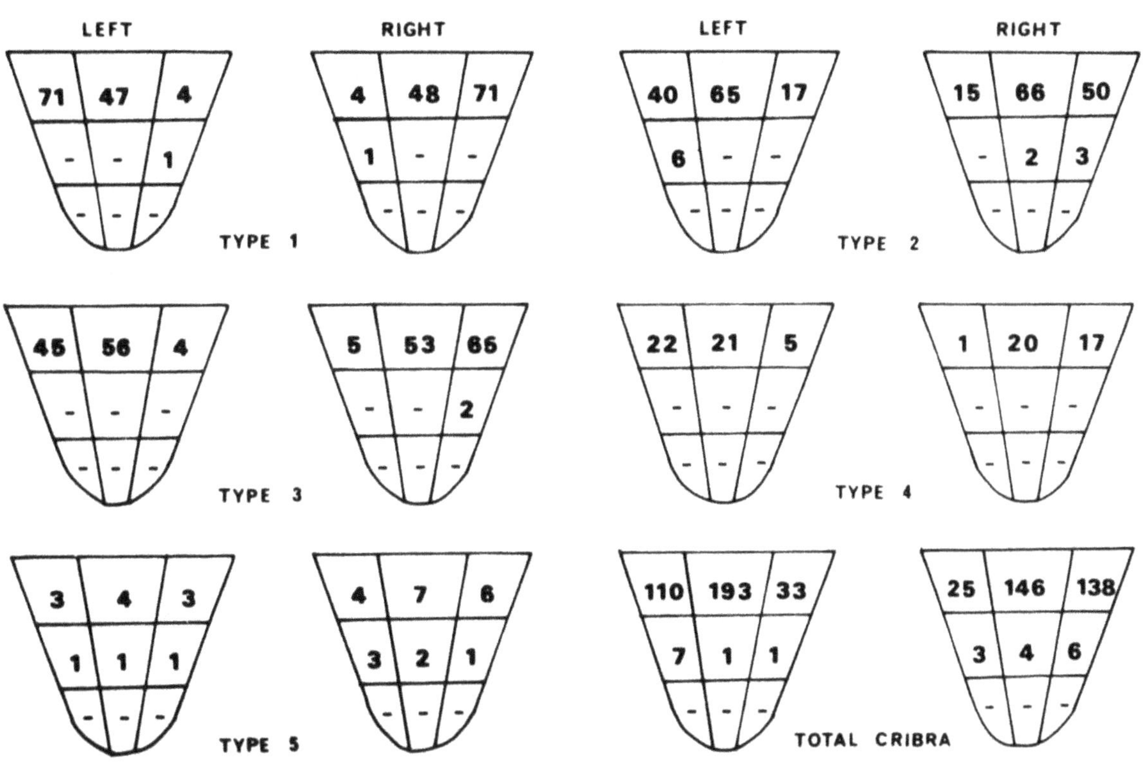

Figure 9.11 Location and type of cribra orbitalia.

10. INTERPRETING MEASURES OF COMMUNITY HEALTH DURING THE LATE PREHISTORIC PERIOD IN MIDDLE TENNESSEE: A BIOCULTURAL APPROACH.

Leslie E. Eisenberg

Interpreting patterns of past human behaviour and understanding the environments in which they developed and changed over time are central concerns to archaeologists the world over. As archaeologists, we commonly study what have been termed "nonintentional and fortuitous records" (Holmes, 1919:1) and, unlike some of our colleagues in anthropology, we cannot enlist the aid of either written records or living informants to help us interpret the remains of past human behaviour. Consequently, it becomes even more critical to examine all possible lines of evidence that may help illuminate the prehistoric past. To this end, multiple strategies are often employed, including the study of artifacts, ecofacts, skeletal remains (both human and non-human), prehistoric disease, settlement patterns, representational art and, even, coprolites.

My research efforts have focused on investigating community health and biocultural adaptation in North American Indian populations inhabiting the southeastern United States during late prehistoric times (A.D. 1250-1450). Collections derived from late nineteenth and early twentieth century excavations and more recent ones in the State of Tennessee have served as the primary source of data, first because of the abundance of late prehistoric burials available for study, and second, because results of this work have the potential to yield new insights into as yet unanswered questions important in understanding archaeology at the regional level in the Southeast.

Prehistoric community health is a topic of interest to archaeologists not only because of the effects disease can have on the ability of a group to adapt successfully to their surrounding biocultural environment, but also because of its direct influence on human behaviour and the resulting reflection of that behaviour in the archaeological record. Since the presence, incidence and patterning of disease is dynamically linked with culture, biology and environment, this study attempts to view disease within a population as a window through which other related aspects of late prehistoric culture may be interpreted. Because of this conceptual and methodological connection which exists, a biocultural approach to the study of disease in prehistory can serve as a sensitive barometer by which changes occurring within the surrounding social and natural environment can be measured.

The study area I will focus on in this paper encompasses the present-day city of Nashville, Tennessee and its environs (Figure 10.1); it is situated within the physiographic area known as the Nashville Basin (Edwards *et al*, 1974:2). The Basin is analytically subdivided into inner and outer sections which are distinguished from each other on the basis of differences in subsurface geology and topography (Braun, 1950:131). Soils in the Nashville Basin, while not uniformly productive, are generally high in silt content, and fertile areas are abundant.

The cultural period under investigation is known as the Mississippian (late prehistoric) period, which, in Tennessee, begins as early as A.D. 900 and continues, in some locations, until approximately A.D. 1550. Temporally, it parallels such Old World events as the Fourth Crusade and the reign of Henry VIII. In comparison with earlier aboriginal North American cultural periods, the Mississippian represents an elaboration, both quantitatively and qualitatively, of many of the social, religious, political and cultural practices known to have been important during preceding times. In general, this period was characterized by cultural stability, and population growth appears to have been one dynamic response to the sedentary occupations of the time. However, for all the advantages it may have conferred, sedentism, often accompanied by population nucleation, carried with it several biological and cultural liabilities, not the least of which were crowding, intra- and inter-group conflict, problems of waste disposal and a decline in overall health (Armelagos and McArdle, 1975).

Towards the end of the Mississippian, but unquestionably prior to either the direct or indirect effects of European contact, a number of significant and unexplained cultural changes are believed to have occurred in portions of the southeastern United States. Of those, one obvious change (as monitored in the archaeological record) involved what appears to be the depopulation of substantial tracts of land which had previously supported thriving human occupations (Williams, 1980). In an attempt to understand this complex demographic process, I turned to the bioarchaeological record to assess whether late prehistoric health stress may have played a part.

The Sample

This paper presents a selected portion of the data obtained from the palaeopathological analysis of the 888 human burials recovered from the Averbuch site, an upland[1] Mississippian village occupied for a minimum of 50 years sometime between A.D. 1275 and 1400[2]. Located at a distance from the floodplain and its highly fertile soils, some nine kilometres north of Nashville, the site plan includes the remains of numerous square houses, three distinct formally structured burial areas and a wooden palisade which encircles a substantial portion of the occupation area (Reed, 1984; Figure 10.2). Averbuch belongs to a local cultural variant known as the Middle Cumberland Culture, whose members share a number of similar traits. Burial in limestone box graves was a characteristic feature of Middle Cumberland mortuary practices. These limestone boxes maintained a basic pH environment, preserving the skeletal remains buried within them; as a result, overall element representation at Averbuch ranges from good to excellent (Figure 10.3).

The overwhelming majority of burials (n = 850, 95.7%) were recovered from the three cemeteries. A small percentage of subadults less than 12 years of age (n = 35, 3.9%), were interred in well-made limestone box or pit graves which were placed beneath the floors of residential structures.

Methodology

In any research endeavour it becomes critical to identify and work within the limitations and constraints posed by the sample itself. For this particular analysis, it is important to recognize that only those pathological conditions that are of a long-standing, chronic nature have the potential to affect bone. Since frequency data for skeletal lesions are not easily translated into the proportion of a past population that may have been ill at any one time, it is more than likely that disease frequencies obtained from this palaeopathological study represent a conservative estimate of the actual number of individuals within the population who were similarly affected.

An objective means for recording palaeopathological changes to bone was developed and, within that framework, each skeleton was examined macroscopically; all bone lesions, their location, distribution and state of healing were coded for later computer analysis and interpretation.

Results of the Analysis

Two distinct types of lesions predominate in the burials from Averbuch. Proliferative lesions of the outer covering of bone, known as the periosteum, are most common in the appendicular skeleton. These lesions develop as an inflammatory response to one or more stimuli, including infection. A second type, represented by resorptive, destructive lesions, is primarily distributed among the bones of the axial skeleton (i.e. the trunk and head), with a particular predilection for the vertebral column.

In order to examine more closely proliferative lesions in the Averbuch series, the tibia was chosen for intensive study not only because it was the long bone most frequently affected by periosteal changes, but also because the presence and degree of those changes have been regarded as a generalized measure of infection in analogous studies of prehistoric community health (Lallo, 1973; Mensforth *et al* 1978, Webb, 1989). The frequency, severity, distribution and often unhealed state of the bony responses observed suggest that infection, rather than trauma (which may have played a part in the initial expression of infection), serves as the most likely explanation for the elevated occurrence of this pathology in the burials (Figure 10.4).

Once the distribution of these lesions throughout the population was examined by state of healing, meaningful demographic and interpretive patterns began to emerge. Interestingly, those individuals less than 2.5 years of age exhibited the greatest frequency of unhealed proliferative tibial lesions, implying that the presence of an active infection just prior to death may have compromised not only normal growth but affected their overall life expectancy as well (Buikstra and Cook, 1981; Figure 10.5). A relative frequency of approximately 13%, observed in those older than 2.5 years, is suggestive of an endemic or chronic level of infection in the population. However, differences in the frequency ratio of active to healed tibial lesions are only statistically significant in those over the age of 15. Additional study is required to determine whether those proliferative tibial changes can be identified with a specific infectious disease syndrome, such as tuberculosis or treponemal infection, or with some other non-specific stressor (Cook, 1976; Ortner and Putschar, 1981:129-133; Powell, 1985, 1988; Steinbock, 1976:137).

---

1. The site is located at an elevation of 165 metres above mean sea level.

2. The Averbuch site was discovered in 1975 during the course of construction for a new housing development. Excavations were originally undertaken by the Tennessee Division of Archaeology and later, on a more massive scale, by the University of Tennessee-Knoxville, Department of Anthropology.

While not observed as commonly as the tibial periostitis, one of the most striking lesions encountered in the Averbuch series is the resorptive, erosive pathology observed in the vertebral column of many of the burials (Figure 10.6). These destructive foci are concentrated primarily in the mid- to lower thoracic and lumbar regions, affecting contiguous body segments and occasionally leading to collapse and ankylosis (fusion) of vertebral elements. The anterior aspect of the vertebral body appears to have been the primary focus, judging from the destructive path that the lesion has followed. Bony bridging between spinal segments is observed with some frequency, and its development is most likely related to the body's attempt to reinforce and stabilize the compromised vertebral area.

Those between 15 and 24.5 years are predominantly affected by vertebral lesions of the thoracic area, although all age groups (with the exception of those older than 50 years) exhibit a statistically significant association with the presence of active lesions. The highest frequency for the distribution of lesions in the lumbar area occurs in individuals over 2.5 years of age, in females between the ages of 15 and 24.5 and in males older than 50 years. In subadults, unhealed lumbar lesions are twelve times more frequent than healed ones. In general, male and female active response frequency for the thoracic and lumbar areas is essentially equal, reflecting their relative representation in the population as a whole.

The age-selective profile for these vertebral lesions in the Averbuch burials suggests that young adults (both male and female) between the ages of 15 and 24.5 and adults older than 50 years of age exhibit the greatest incidence of active pathology. The bodies of the thoraco-lumbar spine appear to be the locus of the osseous infection, although the primary focus was probably pulmonary in origin; involvement of the posterior elements (neural arch and spinous process) is infrequently observed.

While infection is the predominant pathologic response noted in the Averbuch skeletal sample, orbital and cranial lesions (cribra orbitalia and porotic hyperostosis, respectively) have also been observed throughout the population. The presence of these lesions has been hypothetically linked, in other New World prehistoric skeletal series, with iron deficiency anaemia. While many plausible scenarios have been advanced to explain their presence (Angel, 1966, 1967, 1971; Cybulski, 1977; El-Najjar *et al*, 1976; Mensforth, 1985; Mensforth *et al*, 1978; cf. Reinhard, 1988:362-63; Steinbock, 1976; Stuart-Macadam, 1985; Walker, 1986), a most convincing argument has been made by Kent (1986) which suggests that the appearance and distribution of these lesions may be associated, either directly or indirectly, with situations of population aggregation and sedentism.

Interstingly, these lesions appear in the Averbuch sample in spite of the diversity of food resources known to have been available prehistorically in the region, such as elk, white-tailed deer, turkey, migratory waterfowl and hickory nuts (Crites, 1984; Martin, nd; Romanoski, 1984; cf. Martin *et al*, 1985:268). In this late prehistoric cultural context, however, resource availability may not always be synonymous with accessibility; it is highly likely that local sociopolitical conditions during the late prehistoric period may well have limited the Averbuch inhabitants' access to the full range of plant and animal resources known to have been available in the region (Milner *et al*, 1988). Surprisingly, results of recent carbon isotope studies of burials from several Middle Cumberland sites, including Averbuch, reflect a "marked dietary dependence upon maize" (Buikstra *et al*, 1988:248). Given the rich resource base potentially available, this finding was not expected and merits further investigation.

Porotic lesions of the cranial vault were observed in more than one-third (39%) of all individuals from Averbuch with sufficient cranial material to be analyzed (Figure 10.7); of those, 46% were active at death and 54% were healed. Children younger than 2.5 years of age exhibited both active and unhealed lesions, indicating that the stressor or stressors responsible for their occurrence were present prior to, and just after, birth. Some of the most elevated frequencies were observed in individuals between 15 and 35 years of age (Figure 10.8). This finding is intriguing since it has been argued elsewhere that porotic cranial lesions, when observed in adults, simply reflect traces of a remnant childhood anaemic condition (Stuart-Macadam, 1985). Consequently, it is believed by some that its presence in adults may not be a valid means of identifying an exisiting anaemic condition (Huss-Ashmore *et al*, 1982; Stuart-Macadam, 1985; Wing and Brown, 1979). However, in this study, a significant percentage of the porotic lesions in adult crania are clearly unhealed. This may indicate that the skeletal evidence for anaemia may not only be confined to subadults and/or that these individuals were so anaemic that their lesions were not able to heal.

Orbital (cribrotic) lesions, observed in almost 20% of the Averbuch population, are almost equally distributed between adult males and females. They are most often confined to the superior and/or lateral walls of both orbits in both adults and subadults (Figure 10.9). A statistical (Nie *et al*, 1975) cross-tabulation analysis of age, sex and porosity of the orbit indicates that children from 3.0 to 4.5 years exhibit more than one-third of all active lesions observed in the population. That no other age or sex group even approaches these frequencies (except for those aged 5-14.5) suggests that whatever is initiating this response is associated with children just around weaning age and older (Figure 10.10). The only statistically significant difference between active and healed responses in observed in individuals from 15 to 24.5.

While the exact relationship between cranial and orbital lesions continues to be the subject of some debate (Stuart-Macadam, 1989), it has been suggested that because the skeletal appearance of cribrotic orbital lesions usually precedes porotic cranial vault lesions, their primary appearance can be taken as a sign marking the beginning of an iron deficient anaemic condition in an individual (Ortner and Putschar, 1981:259; Steinbock, 1976:242). Parallel frequencies of both orbital and cranial lesions in children between 3.0 and 4.5 years of age suggest that both of these conditons may be responding to similar stimuli; these probably included the effects of weaning and weaning diets, bacterial, viral and diarrhoeal diseases and parasitic infections which are known, in modern Third Worlds contexts, to be particularly devastating in children under five years of age (Kent, 1986). Why the relative frequencies of these bony lesions diverge between the ages of five and 25 in the Averbuch sample is unknown, but it is conceivable that the almost singular presence of cranial lesions after age 25 may be representative of some additional (or more severe) health stress that does not have the capacity to stimulate an orbital response.

The data from Averbuch suggest that the singular manifestation of either porotic cranial or cribrotic oribital lesions alone exceeds any co-occurrence of them and may, therefore, provide a more accurate picture of the generalized distribution of an anaemic condition in a prehistoric population. However, their presence at multiple sites may be a measure of the severity of the stress.

The presence of nutritional deficiencies in the Averbuch population is represented by the cranial and orbital lesions discussed above. While some of these lesions may be due to chronic or seasonal fluctuating levels of iron deficiency anaemia, other contributory factors, such as parasitic infections, food preparation techniques and chronic infections figure prominently in the aetiology of these lesions (Walker, 1986).

Discussion

Specifically, the overall skeletal patterning and sex and age distribution of the lesions related to infection within the Averbuch series permit the development of a differential diagnosis which includes diseases such as tuberculosis, blastomycosis and, possibly, treponemal infection (Buikstra and Cook, 1981; Eisenberg, 1985, 1986a; Kelley and Eisenberg, 1986; M. Powell, 1986, personal communication). The evidence documented for chronic infectious disease stress at Averbuch is consistent with what would be expected in a biocultural setting supporting a high areal population density which, in turn, created the necessary environment for the transmission and maintenance of infectious disease (Kent, 1986; Larsen, 1982; McGrath, 1988; Magennis *et al*, 1985; Mensforth, 1985).

Results of statistical analyses reveal that age, but not sex, is significantly associated with the presence of infection and anaemia at Averbuch. A peak in both conditions at approximately 25 years of age corresponds graphically to a parallel decrease in survivorship during that period as observed inthe mortality profile constructed for the population (Berryman, 1981:60; Figure 10.11). While further study into this relationship may reveal it to be a coincidental one, it is clearly indicative of a period during the Averbuch lifespan when the synergistic effects of such morbid conditions may have overwhelmed the population by lowering the overall level of community health (Scrimshaw, 1975).

It may not be unreasonable to suggest that a number of important subsistence, cultural and defence-related activities may have been affected by the decrease in survivorship among Averbuch adults between the ages of 20 and 30. The loss of a critical sector of economically active adults within the population may have had a significant impact on the timely completion of all aspects of their subsistence round and defence capabilities.

From a biological standpoint, reproductive potential for young adult females must have declined simply because their numbers did as well (Berryman, 1981:71-73; Clarke, 1977:181). An adequate supply of iron would have been most critical for these women, particularly during times of pregnancy and lactation, when maternal mortality would be most vital in maintaining previous levels of fertility, population growth and stability.

Politically, a need for structural defences during the late Mississippian period is demonstrated by archaeological evidence of substantial palisade construction at Averbuch and at other Middle Cumberland villages which may have served as frontier outposts. A qualitative increase in conflict and warfare during the Mississippian, relative to that observed in earlier cultural periods, is also supported by skeletal studies of late prehistoric populations where injuries such as scalping, parry fractures, cut marks and entrance wounds have been recognized; the presence of all of those traumatic conditions at Averbuch, many of which proved to be fatal, underscore the fact that this was indeed a time in prehistory when social stress, often accompanied by physical trauma, was occurring with increasing frequency (Eisenberg, 1986b; Milner *et al*, 1988). Political strain, which appears to have targeted adults between the ages of 20 and 35, was an additional burden on an already anaemic, biologically-stressed and energy-poor population.

Interpretations and Conclusions

In conclusion, the data suggest that the youngest segment of the Averbuch population suffered from infection and nutritional deficiency, along with bacterial, viral and parasitic problems. Many of the individuals who survived the initial disease and stress period in early childhood ran a high risk of dying in young adulthood. Life expectancy at birth for Averbuch males and females was 17.4 and 14.6 years, respectively, among the lowest documented for many comparative historic and prehistoric populations examined (Berryman, 1981:68). If recurrent epidemic episodes of one or more catastrophic events can be postulated to explain the rather unusual bimodal mortality profile obtained (Buikstra and Konigsberg, 1985; Buikstra *et al*, 1988), it may follow that other Middle Cumberland populations, living in marginal situations such as Averbuch, may have been debilitated by heavy and chronic disease loads, and by political and economic factors tied to social predation which may have restricted access to otherwise abundant resources. Clearly, these factors would have affected not only Averbuch subsistence, but political strength and demographic stability as well.

Using Averbuch as a model, this paper underscores the unique contribution that research into prehistoric community health studies can make in understanding elements of past human behaviour that may be difficult to tease out of traditional archaeological research efforts. These palaeopathological data offer one clue for comprehending the complex set of events that may have triggered the phenomenon of depopulation experienced by Middle Cumberland cultures long before European explorers (or their diseases) ever reached middle Tennessee. Additional factors may have included problems related to environmental change, a possible overreliance on selected subsistence resources, such as maize (Buikstra *et al*, 1988), and participation in a demanding tribute system, all of which may have overburdened their production and consumption activities.

In sum, the Averbuch skeletal research provides a baseline from which prehistoric diet, biocultural adaptation and depopulation in middle Tennessee can all be examined productively. Related Middle Cumberland population studies are currently underway and preliminary results suggest a possible link between relative levels of community health and late prehistoric settlement patterning at the regional level. These palaeopathological studies take on an added urgency as many of the existing middle Tennessee skeletal collections may be slated for imminent reburial and would be lost forever to scientific study.

References

Angel, J.L. (1966). Porotic hyperostosis, anemias, malarias and marshes in the prehistoric Eastern Mediterranean. *Science 153*:760-63.

Angel, J.L. (1967). Porotic hyperostosis or osteoporosis symmetrica. In *Diseases in Antiquity* eds. D.R. Brothwell and A.T. Sandison. pp. 379-89. Springfield, Illinois: C.C. Thomas.

Angel, J.L. (1971). *The People of Lerna: An Analysis of a Prehistoric Aegean Population.* Washington, D.C.: Smithsonian Institution Press.

Armelagos, G.J. and McArdle, A. (1975). Population, disease and evolution. In Population Studies in Archaeology and Biological Anthropology: A Symposium. ed. A.C. Swedlund. pp. 1-10. *Society for American Archaeology Memoirs 30.*

Berryman, H.E. (1981). *The Averbuch Skeletal Series: A Study of Biological and Social Stress at a Late Mississippian Period Site from Middle Tennessee.* Ph.D. dissertation, Department of Anthropology, The University of Tennessee, Knoxville. University Microfilms, Ann Arbor, No. 8123141.

Braun, L.E. (1950). *Deciduous Forests of Eastern North America.* Philadelphia: The Blakistan Company.

Buikstra, J.E. and Cook, D.C. (1981). Pre-Columbian tuberculosis in West-Central Illinois: prehistoric disease in biocultural perspective. In Prehistoric Tuberculosis in the Americas. ed. J.E. Buikstra. pp. 115-139. *Scientific Papers 5, Northwestern University Archaeological Program, Evanston, Illinois.*

Buikstra, J.E. and Konigsberg, L.W. (1985). Paleodemography: critiques and controversies. *Am. Anthropol. 87*:316-33.

Buikstra, J.E., Autry, W., Breitburg, E., Eisenberg, L. & Van der Merwe, M. (1988). Diet and health in the Nashville Basin: human adaptation and maize agriculture in Middle Tennessee. In Diet and Subsistence: Current Archaeological Perspectives. eds. B.V. Kennedy and G.M. LeMoine. pp. 243-59. *Proceedings of the Nineteenth Annual Chacmool Conference, Calgary.*

Clarke, S.K. (1977). Mortality trends in prehistoric populations. *Hum. Biol. 49*:181-86.

Cook, D.C. (1976). *Pathologic States and Disease Process in Illinois Woodland Populations: An Epidemiological Approach.* Unpublished Ph.D. dissertation, Department of Anthropology, University of Chicago.

Crites, G.D. (1984). Late Mississippian paleoethnobotany in the Nashville Basin: the evidence from Averbuch. In *Averbuch: A Late Mississippian Manifestation in the Nashville Basin, Vol. 1, Observations.* eds. W.E. Klippel and W.M. Bass. Chapter 12. Report submitted to the National Park Service, Southeast Regional Office, Atlanta, Georgia. Contract No. C-5943(79).

Cybulski, J.S. (1977). Cribra orbitalia, a possible sign of anemia in early historic native populations of the British Columbia Coast. *Am. J. Phys. Anthropol. 47*:31-9.

Edwards, M.J., Elder, J.A. & Springer, M.E. (1974). The Soils of the Nashville Basin. U.S.D.A. Soil Conservation Service, *Bulletin 499.*

Eisenberg, L.E. (1985). *Possible Evidence for Mycotic Disease in a Late Prehistoric Population from Middle Tennessee.* Paper presented at the 12th Annual Meeting of the Paleopathology Association, Knoxville, Tennessee.

Eisenberg, L.E. (1986a). *Adaptation in a "Marginal" Mississippian Population from Middle Tennessee: Biocultural Insights from Paleopathology.* Ph.D. dissertation, Department of Anthropology, New York University. University Microfilms, Ann Arbor, No. 8626867.

Eisenberg, L.E. (1986b). The patterning of trauma at Averbuch: activity levels and conflict during the late Mississippian Period (abstract). *Am. J. Phys. Anthropol. 69*:197.

El-Najjar, M.Y., Ryan, D.J., Turner II, C.G. & Lozoff, B. (1976). The etiology of porotic hyperostosis among the prehistoric and historic Anasazi Indians of the Southwestern United States. *Am. J. Phys. Anthropol. 44*:477-87.

Holmes, W.H. (1919). Handbook of Aboriginal American Antiquities. Bureau of American Ethnology. *Bulletin 60.*

Huss-Ashmore, R., Goodman, A.H. & Armelagos, G.J. (1982). Nutritional inference from paleopathology. In *Advances in Archaeological Method and Theory* Vol. 5. ed. M.B. Schiffer, pp. 395-474. Orlando: Academic Press.

Kelley, M.A. and Eisenberg, L.E. (1986). Blastomycosis and tuberculosis in Early American Indians: a biocultural view. *Midcont. J. Archaeol. 12*:89-116.

Kent, S. (1986). The influence of sedentism and aggregation on porotic hyperostosis and anaemia: a case study. *Man 21*:605-36.

Lallo, J.W. (1973). *The Skeletal Biology of Three Prehistoric American Indian Societies from Dickson Mounds.* Ph.D. dissertation, Department of Anthropology, University of Massachusetts, Amherst. University Microfilms, Ann Arbor, No. 748609.

Larsen, C.S. (1982). The anthropology of St. Catherine's Island 3. Prehistoric human biological adaptation. *American Museum of Natural History, Anthropological Papers 57(3).*

Magennis, A.L., Martin, D.L. & Armelagos, G.J. (1985). Patterns of disease in a marginal environment: the Black Mesa Anasazi. *Am. J. Phys. Anthropol. 66*:201.

Martin, D.L., Goodman, A.H. & Armelagos, G.J. (1985). Skeletal pathologies as indicators of quality and quantity of diet. In *The Analysis of Prehistoric Diets.* eds. R.I. Gilbert, Jr. and J.H. Mielke, pp. 227-79. Orlando: Academic Press.

Martin, T.J. (nd). *The Natural Resources Significant to Prehistoric Cultural Adaptations in the Nashville Basin of Tennessee.* Unpublished paper on file, Department of Anthropology, University of Tennessee, Knoxville.

McGrath, J.W. (1988). Social networks of disease spread in the Lower Illinois Valley: a simulation approach. *Am. J. Phys. Anthropol. 77*:483-96.

Mensforth, R.P. (1985). Relative tibia long bone growth in the Libben and Bt-5 prehistoric skeletal populations. *Am. J. Phys. Anthropol. 68*:247-62.

Mensforth, R.P., Lovejoy, C.O., Lallo, J.W. & Armelagos, G.J. (1978). The role of constitutional factors, diet, and infectious disease in the etiology of porotic hyperostosis and periosteal reactions in prehistoric infants and children. *Med. Anthropol. 2*:1-59.

Milner, G.R., Smith, V.G. & Anderson, E. (1988). *Conflict, Mortality and Community Health in an Illinois Oneota population.* Paper presented at the Centre for Archaeological Investigations Conference, "Between Bands and States: Sedentism, Subsistence and Interaction in Small-Scale Societies", Southern Illinois University, Carbondale, Illinois, April 15-16, 1988.

Nie, N.H., Hull, C.H., Jenkins, J.G., Steinbrenner, K. & Bent, D.H. (1975). *SPSS: Statistical Package for the Social Sciences.* 2nd ed. New York: McGraw Hill.

Ortner, D.J. and Putschar, W.G.J. (1981). Identification of Pathological Conditions in Human Skeletal Remains. *Smithsonian Contributions to Anthropology 28.* Washington, D.C.: Smithsonian Institution.

Powell, M.L. (1985). *Health, Disease and Social Organisation in the Mississippian Community at Moundville.* Ph.D. dissertation, Department of Anthropology, Northwestern University. University Microfilms, Ann Arbor, No. 8600903.

Powell, M.L. (1988). *Status and Health in Prehistory.* Washington, D.C.: Smithsonian Institution Press.

Reed, A. (1984). Site investigations. In *Averbuch: A Late Mississippian Manifestation in the Nashville Basin, Vol. 1, Observations.* eds. W.E. Klippel and W.M. Bass, Chapter 3. Report submitted to the National Park Service, Southeast Regional Office, Atlanta, Georgia. Contract No. C-5943(79).

Reinhard, K.J. (1988). Cultural ecology of prehistoric parasitism on the Colorado Plateau as evidenced by coprology. *Am. J. Phys. Anthropol. 77*:355-66.

Romanoski, S.M. (1984). Analysis of the faunal remains from the Averbuch Site. In *Averbuch: A Late Mississippian Manifestation in the Nashville Basin, Vol. 1, Observations.* eds. W.E. Klippel and W.M. Bass, Chapter 13. Report submitted to the National Park Service, Southeast Regional Office, Atlanta, Georgia. Contract No. C-5943(79).

Scrimshaw, N.S. (1975). Interactions of malnutrition and infection: advances in understanding. In *Protein-Calorie Malnutrition.* ed. R.E. Olsen, pp. 353-68. Orlando: Academic Press.

Steinbock, R.T. (1976). *Paleopathological Diagnosis and Interpretation.* Springfield, Illinois: C.C. Thomas.

Stuart-Macadam, P.L. (1985). Porotic hyperostosis: representative of a childhood condition. *Am. J. Phys. Anthropol. 66*:391-98.

Stuart-Macadam, P.L. (1989). Porotic hyperostosis: relationship between orbital and vault lesions. *Am. J. Phys. Anthropol. 80*:187-93.

Walker, P.L. (1986). Porotic hyperostosis in a marine-dependent California Indian population. *Am. J. Phys. Anthropol. 69*:345-54.

Webb, S. (1989). *Prehistoric Stress in Australian Aborigines.* BAR International Series 490.

Williams, S. (1980). Armorel: A very late phase in the Lower Mississippi Valley. *Southeastern Archaeological Conference Bulletin 22:* 105-110.

Wing, E.S. and Brown, A.B. (1979). *Paleonutrition.* Orlando: Academic Press.

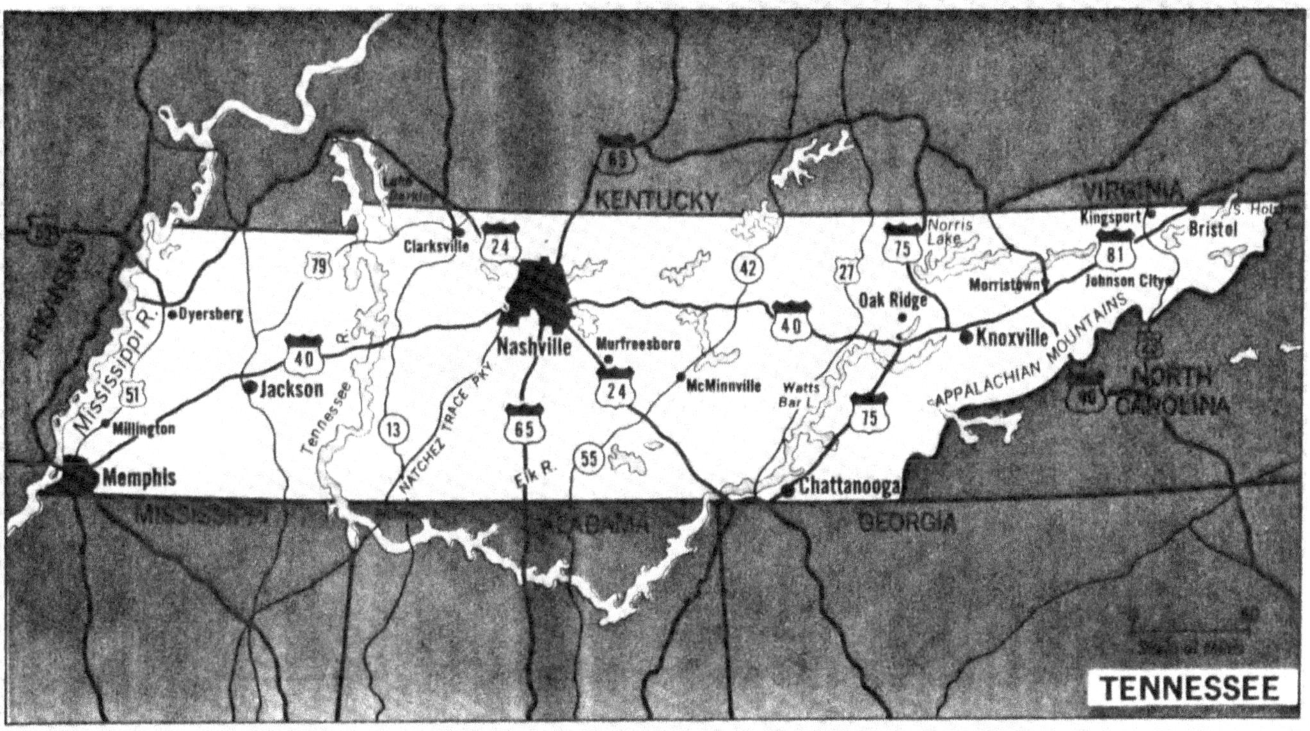

Figure 10.1 Map of the State of Tennessee; the city of Nashville is centrally located in the state and is highlighted in black.

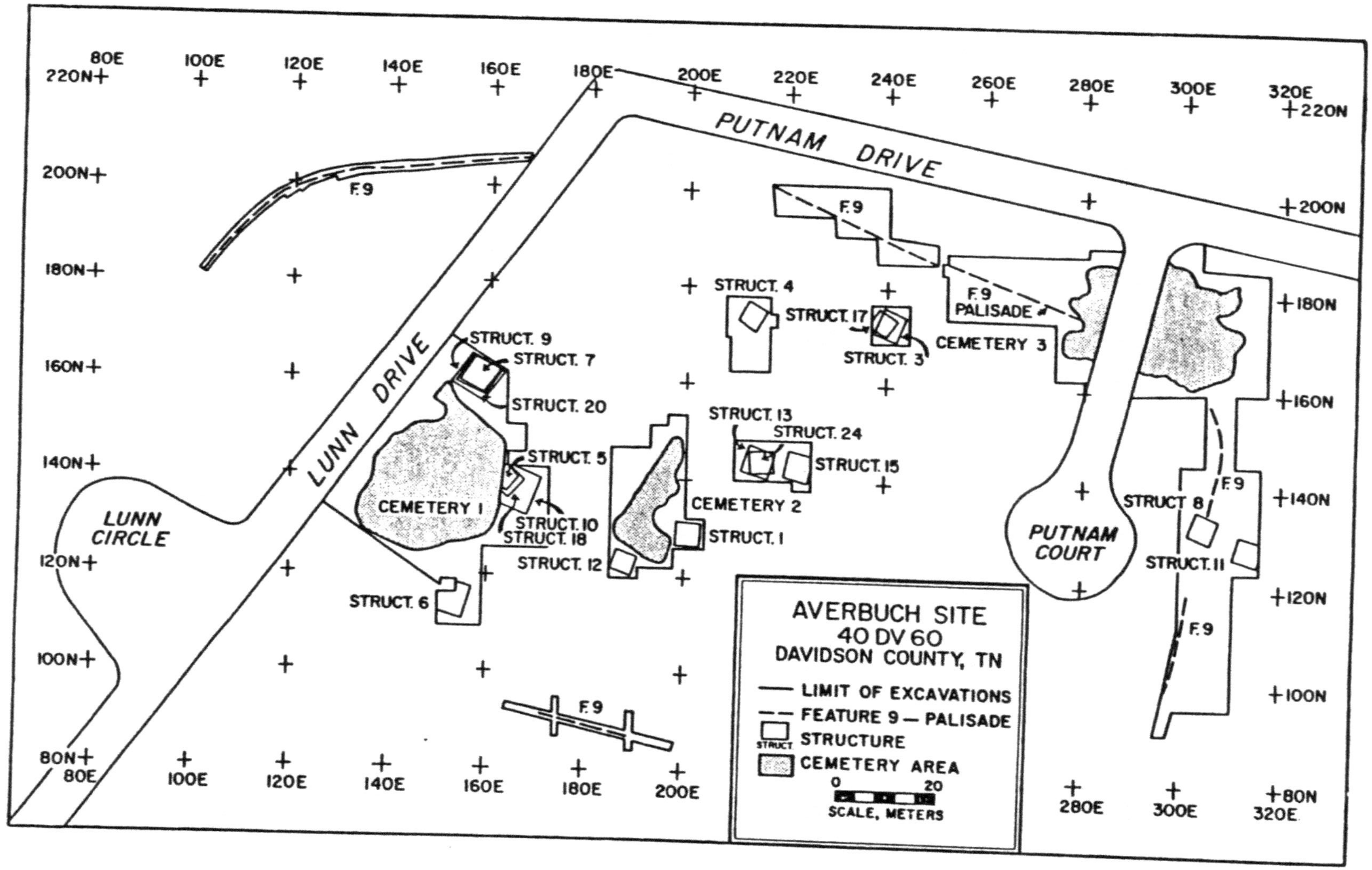

Figure 10.2 Plan view of the Averbuch site, including cemeteries, structures, features and disturbance from modern construction activities (Reed, 1984:8).

Figure 10.3 A typical Middle Cumberland stone box grave. Each grave was constructed of limestone slabs and was sized to fit the individual buried within it.

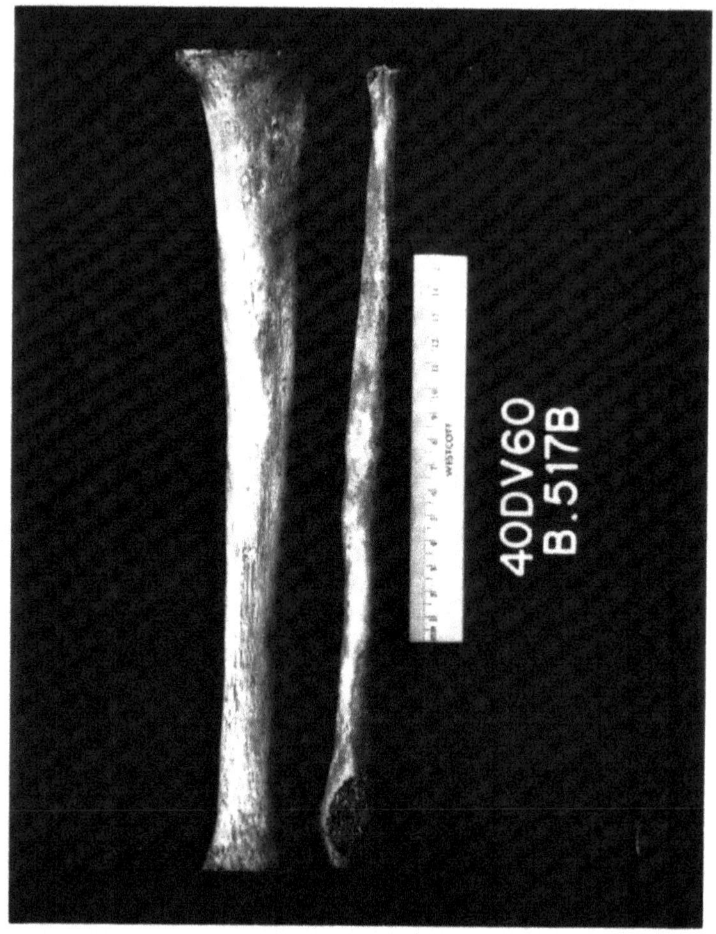

Figure 10.4 Healed periostosis can be observed along the entire shaft of the tibia (left side of photo). Note the irregular surface texture of the cortex of Burial 517B, a female, aged 30-34 years, from the Averbuch site.

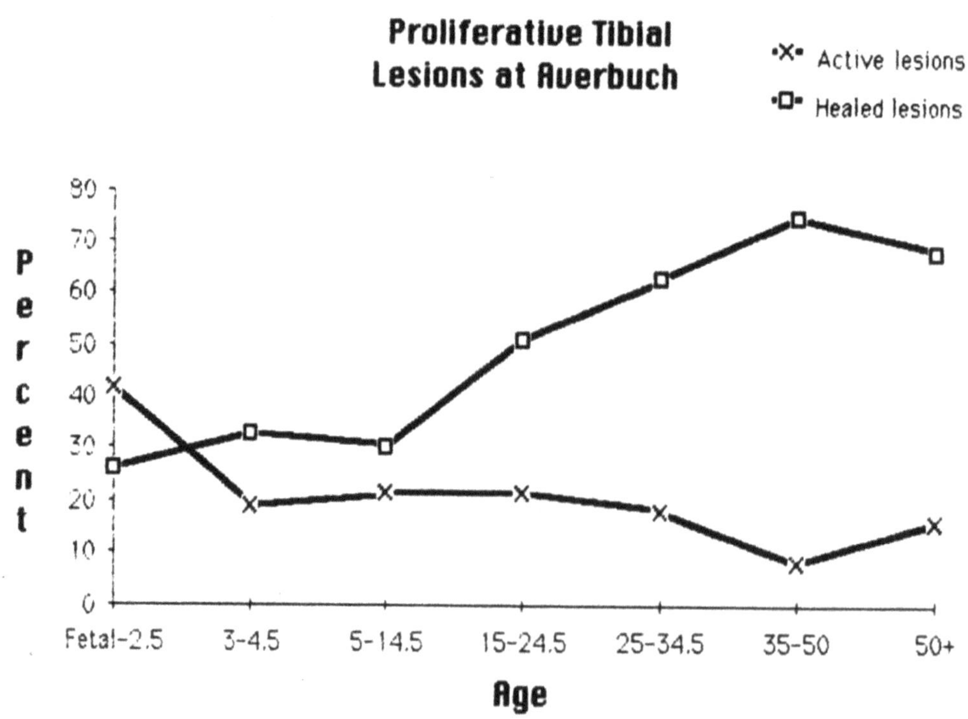

Figure 10.5 Relative frequencies of active and healed proliferative tibial lesions in the Averbuch population. Significant differences in state of healing are observed beginning at age 15 and continue throughout the rest of adult life.

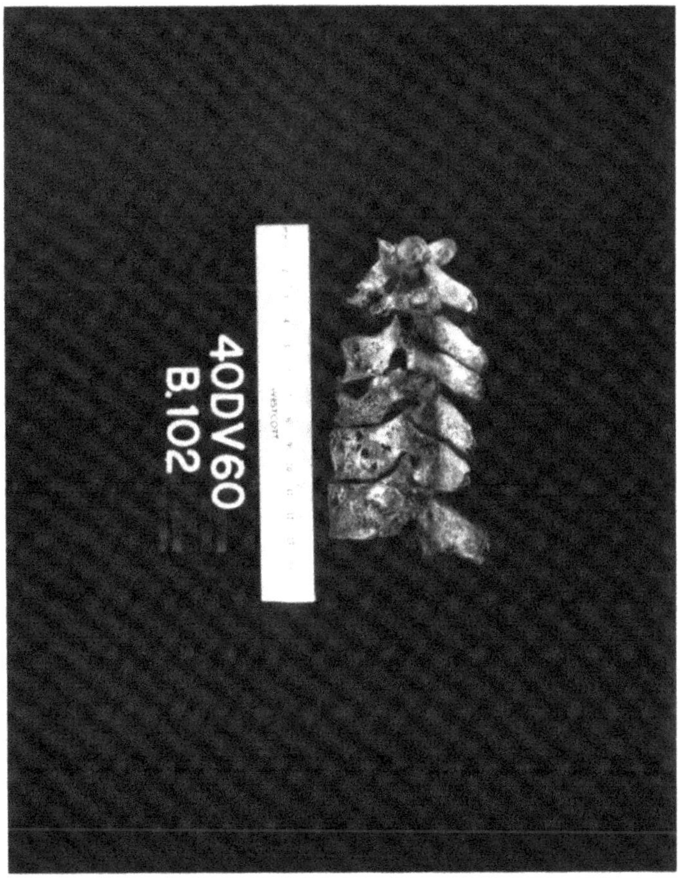

Figure 10.6 Resorptive lesions and collapse of the thoracic spine of an Averbuch female, aged 30-34 years.

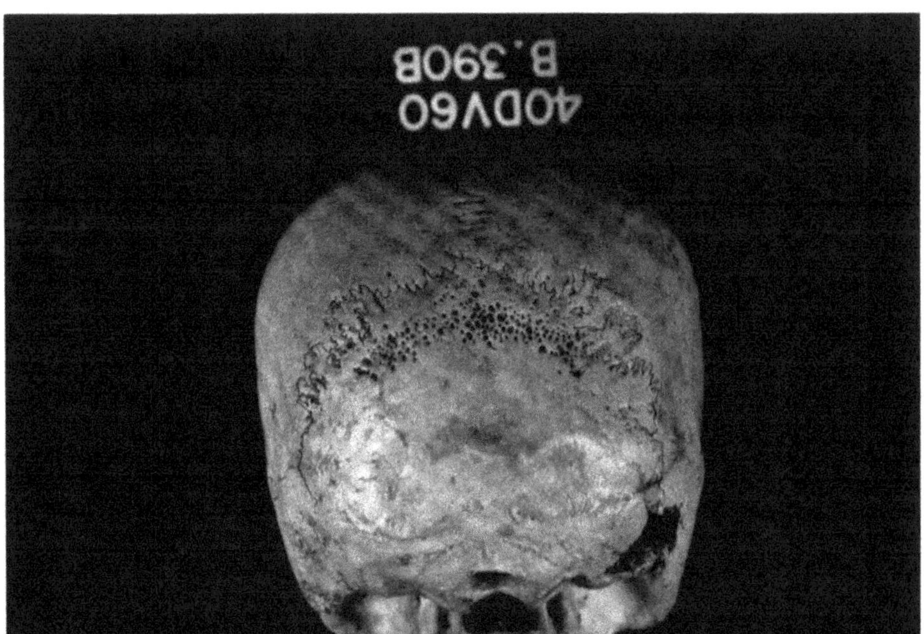

Figure 10.7 Large, active porotic lesions on the posterior of the cranial vault around lambda in an Averbuch female, aged 18.5-19.5.

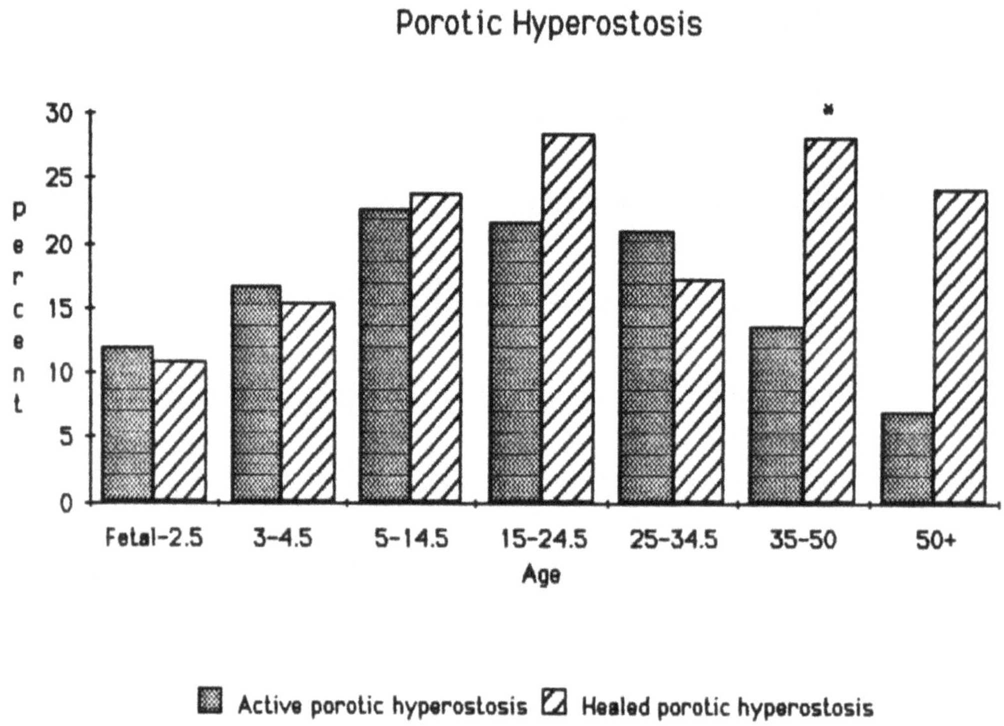

Figure 10.8 Relative percentages of active and healed porotic lesions across all Averbuch age groups. Those from 35-50 years exhibit a statistically significant difference between the two types of lesion (p <.01).

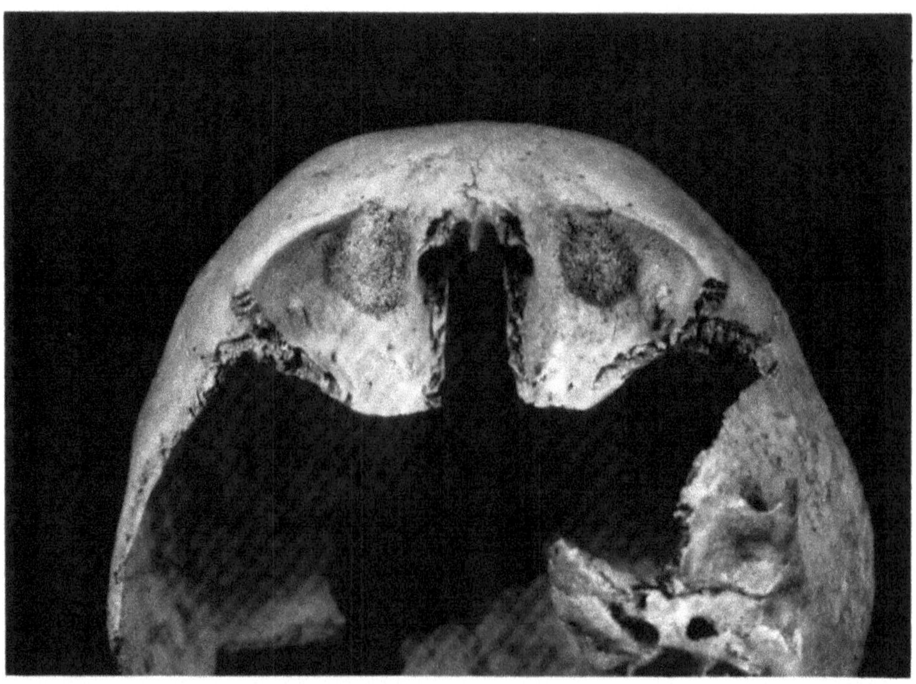

Figure 10.9 Bilateral orbital lesions, which exhibit a combination of healed and active states, in a young child (3.5-4.5 years) from Averbuch.

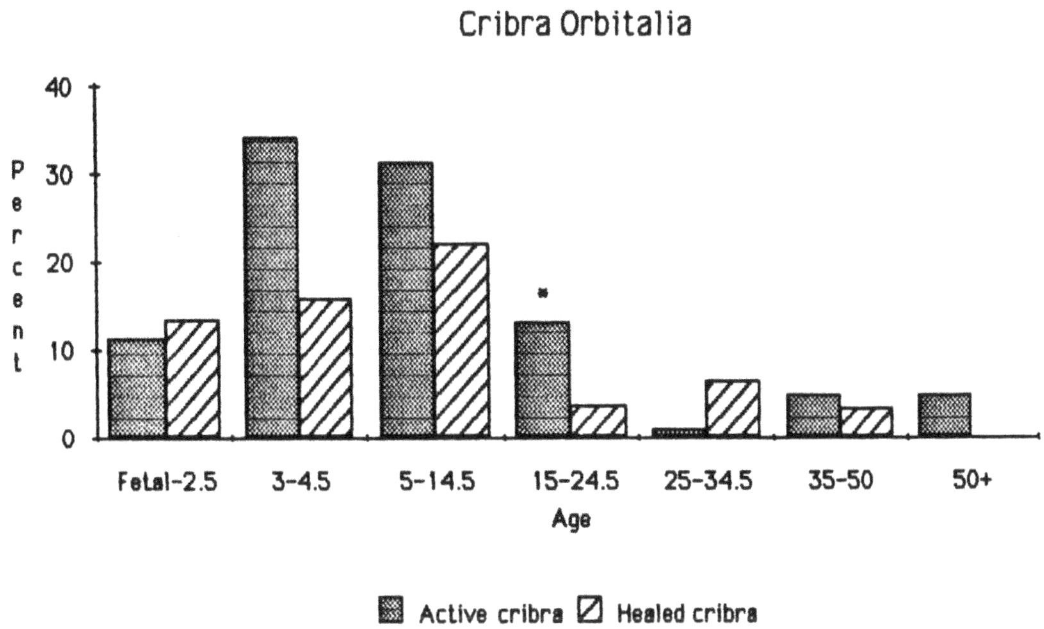

Figure 10.10 Relative percentages of active and healed orbital lesions across all Averbuch age groups.

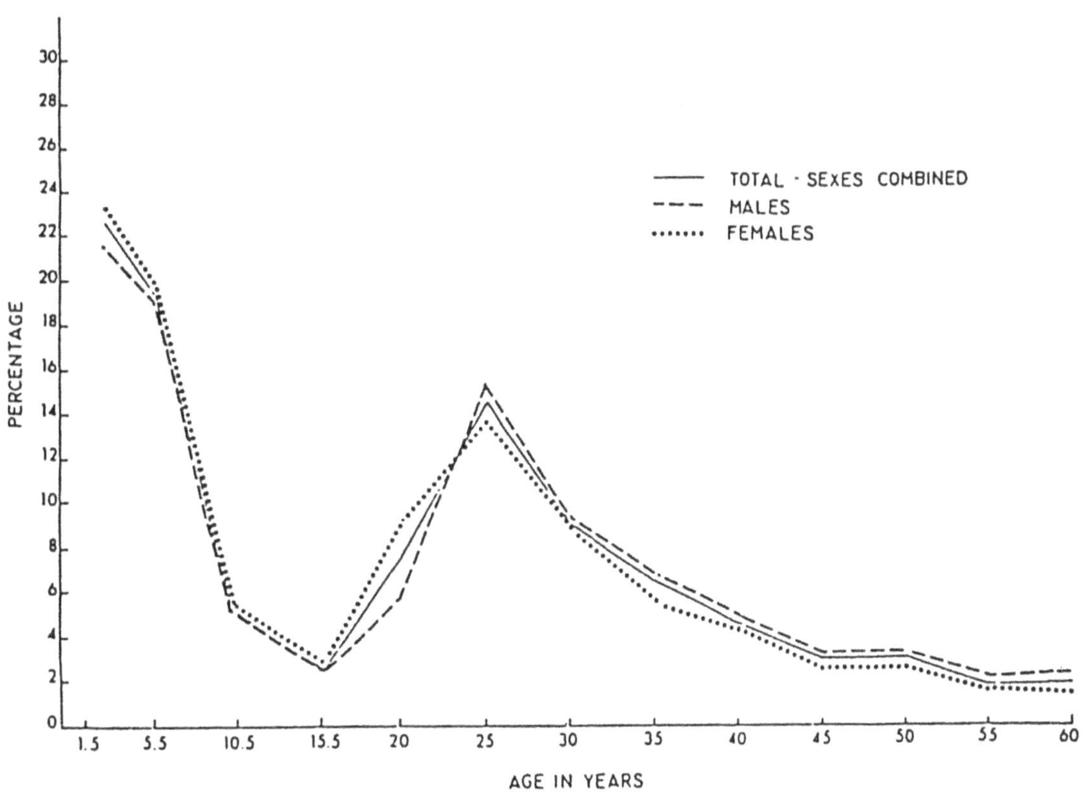

Figure 10.11 Mortality profile of the Averbuch population (Berryman, 1981:60).

11. HEALTH STATUS OF EUROPEAN POPULATIONS AT THE AGRICULTURAL TRANSITION AND THE IMPLICATIONS FOR THE ADOPTION OF FARMING.

Christopher Meiklejohn and Marek Zvelebil

Introduction

The transition from the Mesolithic to the Neolithic is critical to both human biology and archaeology. Many elements taken for granted in human biology are concomitant upon food production, including the modern pattern of human density and population concentration.

In this paper, we examine how the study of past skeletal populations can contribute to our understanding of the Mesolithic-Neolithic transition in Europe. The approach adopted here is biocultural, in which biological and cultural systems are integrated into a single unit, as contrasted with the clinical approach, which is centred on the individual subject. In the latter, the description of the individual becomes the end result of the study. While in the living this provides a basis for treatment, in the archaeological context it can become, in some ways, little more than a taxonomic exercise (see Meiklejohn *et al* (1984) with refererence to this problem in the European Mesolithic and Neolithic).

The biocultural approach recognizes that a diagnosis made on a skeleton is a window to understanding the population from which that individual is drawn. The individual thus becomes representative for the population. The population under study is then the combined and interactive result of both cultural and biological processes.

The Transition to farming in Europe: Some Basic Questions

With a few notable exceptions, the agricultural transition in Europe has had a tradition of being regarded as a replacement of hunter-gatherer Mesolithic cultures by Neolithic farmers (e.g. Lubbock, 1865; Childe, 1928; 1957; Clark, 1952; Piggott, 1954, 1965; Milisauskas, 1978; Ammerman and Cavalli-Sforza 1984; Renfrew, 1987). Within this framework, Neolithic cultures have been viewed as resulting from colonisation by Near Eastern farmers. The role of resident hunter-gatherers has often been disregarded or minimised as a consequence of such views. Yet colonisation models for the origin of farming in Europe entail assumptions which cannot be verified by archaeological data alone. The opposing view, where indigenous adoption of farming has been contemplated for the transition, usually requires explanations concerned with environmental or demographic stress, or social competition.

Some of the key assumptions regarding the Mesolithic-Neolithic transition are:

1) Mesolithic groups were by and large mobile, had low population densities and an egalitarian, "band" level of social organisation: the "free and careless families of bowmen" of Rozoy, (1978).

2) Mesolithic groups had an inherently unpredictable food supply, placing them at risk from dietary stress (see Odell (1974) for a discussion of the widespread nature of such an assumption).

3) Neolithic groups were more sedentary. This removed constraints imposed by mobility on population growth and increased the reproductive capacity of such groups, leading to higher population densities and more stratified social organisation (Bender, 1978; Ammerman and Cavalli-Sforza, 1984; Dennell, 1983).

4) Agro-pastoral farming increased the productivity, predictability and quality of the food supply, allowing Neolithic populations to sustain rapid population growth (Ammerman and Cavalli-Sforza, 1984; Renfrew, 1987).

5) Neolithic groups colonised hitherto unfarmed areas, absorbing or killing off scant Mesolithic populations (Ammerman and Cavalli-Sforza, 1984; Vencl, 1986).

Although all of these assumptions have been examined on archaeological grounds, the evidence is often tenuous and subject to contrary interpretations; the study of human remains within the biocultural framework provides convergent information which is directly related to the biological and ecological conditions of the people involved.

The type of questions that we would like to direct towards human biological evidence, are, for example:

1) What is the relative health status of Mesolithic and Neolithic populations, and what are the patterns of stress?

2) What evidence is there for particular stress conditions?

3) What evidence persists in skeletal remains for the change from a foraging to a farming diet?

4) What evidence is there for social differentiation in dietary patterns ?

5) What effects does the agricultural transition have on the health status of Neolithic populations?

6) What evidence do we have for population movements at the time of the transition?

The Biocultural Evidence

This paper cannot be an exhaustive survey of all of the above questions and of the further related issues that arise from them. By selecting some examples, however, we hope to illustrate how combining cultural and human biological lines of evidence can shed light on some of the questions asked. For the sake of brevity, we have ordered these into a number of propositions:

Proposition 1: In general, there are few signs of poor health or dietary stress among the Mesolithic populations. Rates of pathology for the European Mesolithic are not particularly high.

For the Mesolithic, reported postcranial pathologies have primarily been degenerative conditions, most notably arthritis (Meiklejohn *et al*, 1984). Areas most commonly afflicted are the vertebral column (>50% of reported cases), the shoulder and upper arm, and the foot. All of these are suggestive of activity induced degenerative disease. For the succeeding Neolithic, the incidence of arthritis is reduced, compatible with worldwide figures for arthritis noted by Cohen and Armelagos (1984a). This suggests a decrease in overall workload. Relatively little other postcranial pathology occurs within the Mesolithic. With the possible exception of the Iron Gates site (Garashanin, 1978; Žinanovic, 1975), the limited evidence for infectious disease suggests a lack of the infectious "reservoir" that is so typical of urbanised conditions (see Anderson and May, 1979; May and Anderson, 1979 for a generalized discussion of the dynamics of infectious disease).

At first, cranial pathology would also appear infrequent in both Mesolithic and Neolithic Europe. Recent evidence, however, suggests the emergence of a dual pattern of pathology in the Mesolithic. One area of cranial pathology with apparent higher incidence than has been previously reported is cranial porotic hyperostosis. Previous work presented only two examples (Cuzoul de Gramat, France, and Falkensteinhohle, Germany [Meiklejohn *et al*, 1984]). However, recent work with Danish samples, primarily (though not entirely) in Ertebølle contexts, suggests that it is not uncommon (e.g. Bøgebakken, Bergmansdal, Bloksbjerg and Lundby). The presence of porotic hyperostosis has also been noted at Franchthi Cave (Angel, 1984). While no porotic hyperostosis has been noted at Vlasac ([Iron Gates of the Danube] Nemeskeri and Lengyel, 1978), Angel (1984) implies that it was present. Porotic hyperostosis is secondary to reduced levels of iron in the blood supply, thus leading to increased red cell production in the diploe of the cranial bones (among others) (Goodman *et al*, 1984b). The most common explanation is iron deficiency anaemia (Hengen, 1971). This may be expanded to include malarially induced anaemia, and other "haemoglobin derived" anaemias (Goodman *et al*, 1984b), and nutritional stress. Studies by Lallo *et al* (1977) and Mensforth *et al* (1978) further point to infectious disease as a conditioning factor, while Kent (1986) links the high incidence of hyperostosis among the Anasazi to sedentism, higher population density and density-related diseases.

Hyperostosis resulting from nutritional stress occurs mainly among populations that are heavily dependant on cereal diets. In the case of the Ertebølle culture, dental pathology, chemical analysis of bone, and faunal remains all indicate a meat-based and marine diet (Rowley-Conwy, 1983; Tauber, 1981; Price, 1989).

However, the new cases of porotic hyperostosis noted here do not fit the classic pattern suggested above. In this case, infection and/or increased sedentism are more likely causes of pathology than is iron deficiency or any other similar anaemia. A number of Mesolithic crania from Denmark show very thick cranial vaults, and infestation with various fish tapeworms is a possible explanation (J. Balslev Jørgensen and E. Brinch Petersen, pers. comm.).

Archaeologically, we know that salmon, cod, eel and marine mammals were common sources of food in the circum-Baltic environment (Rowley-Conwy, 1983; Andersen, 1987; Rowley-Conwy and Zvelebil, 1989). Also, as noted, hyperostosis may have been present at Vlasac, while at Franchthi, 50% of the individuals suffered from the condition (Angel, 1984). Catfish, in the case of the former, and tuna, in the case of the latter, are found in high quantity among the respective faunal remains. Moreover, at Vlasac, the skeletal population showed a significant incidence of osteomalacia (15% of males, 25% of females) and rickets (44% of child skeletons)(Nemeskeri and Lengyel, 1978; y'Edynak, 1989), both indicative of the presence of vitamin D and/or calcium deficiency. This is a suprising condition in a population heavily dependant on fish, since fish has a good supply of both of these nutrients (Printz, 1987).

Although it may be possible that the cause of hyperostosis lies in malarially induced anaemia we suggest that this is unlikely. We would argue that the best explanation for the co-occurence of these insults as a group appears to be parasitic infestation (for a discussion of the effects of such parasitism see such sources as Marquardt and Demaree [1985] or Bogitsh and Cheng [1990]). The presence of such parasitism and disease would automatically result in impairment of the function of both the kidneys and the intestinal tract. This would be the result of a reduced ability of the liver to convert vitamin D into an active agent. The immediate effect would be the loss of vitamin D and calcium from the intestine during normal elimination. The end result, the secondary loss of calcium through the kidney and its absence for resorption in the intestine, would aggravate the general condition of parasitism. If the implicated tapeworm is *Diphyllobothrium*, which has the dog as an important intermediate host, then a further effect is Vitamin B12 deficiency, which leads to lowered haemoglobin production, and therefore to anaemia (Marquardt and Demaree, 1985; Bogitsh and Cheng, 1990). *Diphyllobothrium* is a fresh -water parasite, found in, among others, pike, salmon and trout. Scandinavia and the Danube basin are both areas where this parasite is endemic. Thus, significant levels of intestinal parasites, among fishing communities of northern and eastern Europe, may have compromised an otherwise high quality diet, in terms of benefit to the individual.

We would also like to point out a further possible mechanism for the appearance of bone stress markers in the presence of a high marine diet. Lazenby and McCormack (1985) point out that high usage of salmon can result in hypervitaminosis-D. Since vitamin D can stimulate calcium withdrawal from the skeleton, this provides a basis for markers such as porotic hyperostosis. It should also be pointed out that hypervitaminosis is a potentially fatal syndrome, and that the situation outlined by Lazenby and McCormack is thus an extension of the potential for protein poisoning that can result from excessive use of marine resources (Noli and Avery, 1988).

A similar pattern of insults may have occurred as a result of higher population density and increased sedentism of these populations, if hyperostosis is taken as an indicator of higher rate of infection (i.e. Kent, 1986). There is strong cultural evidence, in terms of seasonal indicators of occupation and the permanent nature of dwellings, that both the inhabitants of the Iron Gates and the populations of the Ertebølle were more sedentary than other Mesolithic groups (Rowley-Conwy, 1983; Price, 1985; Printz, 1987; Clason, 1980; Bolomley, 1973; Voytek and Tringham, 1989; Chapman, 1989, etc), although we recognize, in this context, that sedentism does not necessarily mean year-round permanent habitation of a single site.

Sub-optimal levels of health can be also detected through the presence of dental enamel hypoplasia. Previous work had suggested that the trait was almost universally absent in the European Mesolithic and Neolithic (Meiklejohn *et al*, 1984). More recent work on both Danish and Portugese collections indicates that low levels of enamel hypoplasia are present in both cases (Alexandersen, 1988; Lubell and Jackes, 1989; Lubell *et al*, 1989), although only a few cases are of the severe type reported in some areas of North America, for example. In contrast, y'Edynak (1989) reports the presence of enamel hypoplasia on 70% of all teeth at Vlasac. The levels recorded for the second molars are taken as an indication of the severity of the condition. Hypoplasia is also present at Padina (y'Edynak, 1978). In general, studies of enamel hypoplasia levels across the Mesolithic-Neolithic transition have not been carried out. One case, however, is that of Alexandersen for southern Scandianavia (1988). In this case he has shown that 53.5% of the individuals from Vedbaek-Bøgebakken, Skateholm I and Skateholm II show mild hypoplasia (n=58). In comparison, 45% of Danish middle and late Neolithic individuals show the trait (n=102). The difference is not significant (chi-square=1.03, p=.3095). It should be stressed, however, that the dietary base of these two samples is profoundly different, as shown by bone isotope studies (Tauber, 1981; Price, 1989). For this reason we cannot assume that the cause of the hypoplasia is identical in the two cases. Unfortunately, a similar enumeration of enamel hypoplasia for other regions in the Neolithic has not yet been carried out, making it impossible at this time to make generalizations about the incidence of hypoplasia across the agricultural transition.

The above discussion suggests a dual pattern in the patterns of these pathological conditions in Europe: both porotic hyperostosis and enamel hypoplasia are more common than has been believed, but occur principally on settlements in eastern and northern Europe. Here they are associated strongly with fishing and incresed residential permanence, thus leading us to a second proposition:

Proposition 2: There is evidence for spatial variation and major dietary differences in late Mesolithic populations prior to the introduction of farming.

This proposition can be best illustrated by dental pathology. The most widely reported dental pathology in Europe has been caries. In the earlier survey (Meiklejohn *et al*, 1984) Mesolithic caries levels were reported as low compared to Neolithic levels (1.9% vs 4.2%). However, more recent work suggests a more complex pattern, including a lack of geographic uniformity (Meiklejohn *et al*, 1988; Frayer, 1989). Current overall results are presented in Table 1. The earlier reported figures clearly underreported caries rates for the Mediterranean region. The overall pattern is clear, with average caries rates increasing from the Baltic through mid-latitude Europe to the Mediterranean basin. The exception to this pattern is once again at Vlasac and Padina in the Danube Gorges, where absence of caries (Frayer, 1989; Žinanovic, 1975) aligns this settlement with the Baltic sites rather than to mid-latitude sites such as Teviec and Hoedic. The differences are highly significant. What do they suggest?

Critical to caries development is carbohydrate consumption. Dietary carbohydrate affects plaque pH, increasing acidity to the point where dental enamel is dissolved. Dietary protein results in increased alkalinity of the plaque, leading to mineral crystalisation and calculus. The presence of calculus and caries indicate alternate periods of acidity and alkalinity and the alternation of carbohydrate and protein dietary sources. (For more comprehensive reviews of caries formation, see Hillson (1986), Meiklejohn *et al* (1988) and Frayer (1989) and their contained sources).

From the above, it can be seen that the incidence of caries is in inverse relationship to the bimodal pattern of other pathologies, noted above. The data from Vlasac, Padina and the Scandinavian materials show calculus, but caries are virtually absent. A protein-based diet is supported, a conclusion bolstered by faunal analysis for the two areas, and bone chemical analysis for the Baltic sites. As shown above, these sites also have a higher incidence of other pathology.

In the Mediterranean samples, both calculus and caries are present, suggesting a mixed protein and carbohydrate diet. Again, this is supported by palaeoeconomic and bone chemical analysis at sites such as Uzzo, where trace element analyses generally suggest a carnivorous diet and particularly high consumption of molluscs, while relatively high incidence of caries and the presence of macrobotanical remains indicate a dietery component of fruits, nuts and berries (Meiklejohn *et al*, 1988; Borgognini-Tarli, 1989; Frayer, 1989). Macroscopic finds of legumes, nuts and fruits were also made at Balma Abeurador in France (Vaquer *et al*, 1986), Fontbregoua (Courtin, 1974) and elsewhere in the Western Mediterranean (Binder, 1989) as well as at Franchthi in Greece (Jacobsen, 1976; Hansen and Renfrew, 1978). At Cingle Vermell, in Catalonia, where eight out of nine teeth were found to be carious, the associated cultural layer contained the remains of walnut, plums, cherries, pine nuts and acorns (Turbon, 1989). There is also evidence for the use of hazel nuts, cereals and Chenopodia at Vlasac (Carcuimaru, 1978; Printz, 1987), though without caries, and at Icoana (Carcuimaru, 1978; Voytek and Tringham, 1989). The suggestion of an increased carbohydrate intake is further supported by delta 13C figures from Portugal (Moita do Sebestiao, -16.4; Cabeco da Arruda, -17.6; Neolithic - 19.8)(Lubell and Jackes, 1989).

The incidence of caries in the West Mediterranean Late Mesolithic - 9.45% - is about 4 times higher than the caries rate among the Natufians of 2-3% (Smith *et al*, 1984; Smith, 1989) and about twice as high as the comparable figure for reported levels from the European Neolithic. It is also higher than Neolithic levels in the Levant (Frayer, 1989; Meiklejohn *et al*, 1988; Smith *et al*, 1984) and can be compared to the levels attained in the carbohydrate oriented agricultural communities in North America during the Mississipian and Woodland periods (Cook, 1984; Perzigian *et al*, 1984; Rose *et al*, 1984).

Complementary data come from an examination of tooth size in the Portugese Mesolithic sites of Moita do Sebastiao and Cabeco da Arruda. A general decrease in size was noted between the sites, a period of only half a millennium. This trend seems to continue into the Neolithic (Meiklejohn and Schentag, 1988). However, these changes began within the Mesolithic and are therefore not the product of the agricultural transition per se. A similar pattern has been observed at the Yugoslavian Iron Gate sites, again within the Mesolithic (y'Edynak, 1989). It has been suggested that the reduction in posterior tooth size, and in the dentition in general, reflects a shift to a more carbohydrate diet, usually viewed as the product of agriculture (Calcagno and Gibson, 1988). However, a significant shift to any softer diet could lead to a decrease in mandibular size and in dentition (y'Edynak, 1978, 1989). Bearing in mind associated evidence, dental reduction supports the case for a shift from meat-based to either the fish-oriented or carbohydrate-oriented diet in the late Mesolithic. Thus a dietary shift can be posited from evidence from both dental caries and dental size (though Frayer (1989) argues against an overall temporal trend in caries rates within the Mesolithic).

Based on these considerations, a dual pattern of diet and lifestyle seems to be emerging for the European Mesolithic (Fig. 1). In the west Mediterranean, a case can be made for the consumption of a balanced diet, based on plant foods, fish and meat. The greater importance of a carbohydrate diet in the Mesolithic gives some support for the existence of indigenous plant husbandry, or at least a shift towards more intensive gathering of wild plant foods (though the evidence from Portugal is for the later Mesolithic, the data from Uzzo and Molara in Sicily date to before 7000 bc (Borgognini-Tarli *et al*, 1989; Constantini, 1989 and contained references). Such a pattern has been described by Cohen (1977, 1989) and Binford (1968, 1983) as a shift from preferred to sub-optimal resources, and therefore as an indication of population pressure and demographic stress. And yet, with the exception of caries, the indications of physical stress are lower in this part of Europe than elsewhere.

In the south Baltic and at Lepenski Vir, another dietery pattern is emerging (Fig. 1), which is based on the great significance of aquatic resources. Faunal and technological evidence from other sites in northern and eastern Europe suggest that this has been the prevalent pattern over much of this area (Clark, 1976, 1952; Loze, 1979; Tringham, 1973; Zvelebil, 1986a,b;

Rowley-Conwy and Zvelebil, 1989). Human skeletal analysis suggests that this "north-east" European pattern was associated with greater levels of pathology and physical stress than was the case with the "West Mediterranean" pattern. The pathology may be reflective of increased seasonal variation in the food supply, increased residential permanence, and/or presence of parasitic disease arising from consumption of contaminated fish.

Proposition 3: We suggest that on the present evidence, little or no difference exists in the overall health status of Mesolithic and Neolithic populations.

Admittedly, this is a difficult proposition to evaluate at present, because of wide discrepancies in the approach, the level of analysis and the size and the distribution of skeletal samples representing Mesolithic and Neolithic populations. At present, there has been more extensive analysis of Mesolithic than of Neolithic remains in a manner allowing for an accurate reconstruction of health status. Meiklejohn *et al* (1984) have discussed the problem of deriving such information from the Neolithic data available. This appears to be a reflection of the clinical approach to the description of human remains (as discussed above). It also is, no doubt, a reflection of the traditional view, where stress was not expected among farmers, whereas, among foragers, dietary stress was anticipated. This view fits with either of two models, the classic "poor starving savages" caricature of pre-Neolithic life, or with stress as the "prime mover" of the transition to farming as per Cohen (1977). The end result is that we can only summarise the current state of the evidence, the comparative value of which is limited.

For example, the present evidence shows higher levels of arthritis, enamel hypoplasia and porotic hyperostosis in the Mesolithic, although the latter two might have been under- reported in the Neolithic samples (Meiklejohn *et al*, 1984). Y'Edynak (1978) notes that enamel hypoplasia at Padina (Iron Gates) was more common in Mesolithic than in Neolithic levels. Pap (1989) reports the presence of porotic hyperostosis in the early and late Neolithic of Hungary, while noting the absence of enamel hypoplasia. Many other reports do not mention either condition (Meiklejohn *et al*, 1984).

Similarly, it was thought that caries frequencies were higher in the Neolithic than the Mesolithic (see above). However, the situation is far more complex. As a simple example, the recalculated overall Mesolithic caries rate from Table 11.1 is 5.03%, higher than the previously noted figure of 4.2%, derived from 16 Neolithic sites from the Netherlands, Belgium, France and Portugal (see above, and appendix to Meiklejohn *et al*, 1984). Stature is clearly a marker for the overall health and nutritional status of a population (see Goodman *et al*, 1984b, this volume). Stature estimations for European samples from the Upper Palaeolithic to the Neolithic have been reported by Meiklejohn *et al* (1984). From the Upper Palaeolithic until the Neolithic, populations become progressively shorter. Only after the late Neolithic is the trend reversed. The difference between the average Mesolithic and average Neolithic stature does not reflect abrupt change. In fact, overall Mesolithic (n=82) and Neolithic (n=190) sample means are strikingly similar; 167.7 and 167.3 cm for males, 155.6 and 154.1 cm for females. Whatever is producing short stature in the Mesolithic is still a factor in the Neolithic. If reduction in stature indicates nutritional stress, as some have claimed, then the stress was not resolved until the end of the Neolithic. This would support evidence that, while the adoption/spread of farming economy may be, in part, a response to pressures set up in late Mesolithic times, it does not fully alleviate them (see Cohen and Armelagos, 1984a; Roosevelt, 1984). Farming may simply allow more people to survive without improving their general health.

Proposition 4: there is evidence for increase in population density in some areas of late Mesolithic Europe, particularly in Northern and Eastern Europe.

In areas such as southern Scandinavia or the north-east Baltic, and in the North European Plain, the cultural evidence consists of an increase in the size and permanence of settlements, decrease in the size of apparent territories and the appearance of cemetaries (Fig. 1) (Brinch Petersen, 1973; Champion *et al*, 1984; Newell, 1984; Constandse-Westermann *et al*, 1984; Zvelebil and Rowley-Conwy, 1986). These cultural developments appear to be reflected in the biological record, although the evidence remains tenuous. The territorial behaviour and density increase are mirrored in the number of skeletons showing signs of violence. Types of injury include apparent blows to the skull, possibly caused by heavy striking weapons, and embedded projectile points, most frequently in the vertebral column (Brinch Petersen, 1981; Newell, 1984; Constandse-Westermann and Newell, 1984; Meiklejohn *et al*, 1984; Angel, 1984; Printz, 1987). If such behaviour is density dependant, as has been suggested by workers such as Chagnon ((1972), see also Hyman and Scutt, this volume), then we have a biological marker for this demographic shift. As noted above, density increase may also be implicated in the presence of conditions such as porotic hyperostosis and, possibly, enamel hypoplasia.

Another trend suggestive of increased density during the late Mesolithic is the increased variability and greater circumscription in physical characteristics in these populations. This may suggest that a shift has taken place from "band level" societies with open mating networks (Meiklejohn, 1974, 1977) to more circumscribed societies with greater genetic isolation (Newell, 1984; Constandse-Westermann *et al*, 1984). Finally, an analysis of the distribution of ornaments in European sites suggests an increased circumscription and integration of cultural units over the duration of the Mesolithic, occurring earliest in the south and south-east (Constandse-Westermann & Newell, 1988).

However interpreted, this evidence at the very least points to regional variability in population numbers during the late Mesolithic and implies that at least in some areas, population density may have been as high as it was in the succeeding early Neolithic.

Proposition 5: There is evidence for differences in health and social status among the inhabitants of Mesolithic settlements.

The human biological evidence points to differences in health status and pattern of insults between males and females, which, although contingent to some extent on the sex of individuals and the childbearing role of women (see Bush, this volume), may additionally indicate an unequal access to resources, which, on the whole, prejudiced the health status of females. Females appear to suffer from greater incidence of dental disease and nutrient deficiency. Frayer (1989:262) noted the greater incidence of caries among females:

> "Female caries rates are about double those for the males and significantly different at the 0.007 level or better".

At Vlasac, females suffered from more severe levels of alveolar resorption and osteomalacia, though males had a higher incidence of calculus (y'Edynak, 1989). Both of these data sets indicate a greater lack of calcium, protein and possibly other nutrients among females, which could be caused, at least in part, by pregnancy and lactation. The question is whether these differences can be explained by childbearing alone. The differential pattern in the incidence of caries and calculus suggests that a dietary factor may have been additionally implicated: males consumed more protein and calcium, while females concentrated more on a carbohyrate diet (y'Edynak, 1989). This pattern finds some support in what we know of food sharing practices and food procurement among modern hunting and gathering people, where males may be able to capitalize on the consumption of the most nutritious meat parts (Speth, 1989, 1990; Kent, 1989). A comprehensive trace element analysis of skeletal samples should resolve this question.

Cultural evidence for the existence of social ranking is based mainly on the examination of mortuary contents from Mesolithic cemetaries. The existence of generalised wealth or status ranks (O'Shea and Zvelebil, 1984; Clark and Neeley, 1987; Constandse-Westermann and Newell, 1989), of ascribed status (O'Shea and Zvelebil, 1984; Clark and Neeley, 1987) and of special, occupationally-related statuses (O'Shea and Zvelebil, 1984) has been noted. In general it seems that both men and women could acquire high rank in Mesolithic society (O'Shea and Zvelebil, 1984; Constandse-Westermann and Newell, 1989) although men tended to aquire higher rank more often then women (O'Shea and Zvelebil, 1984). Constandse-Westerman and Newell (1989) examined skeletal markers of mechanical stress in the light of the social differences noted above and found that higher ranking males show stronger limb lateralisation than those in lower ranks, while higher ranking females show less limb lateralisation and arthritis than lower ranks. The meaning of these differences remains unclear; one possible explanation may be that males acquired status through specialised physical activity, while high female status was ascribed by descent, by association with dominant males, or through activity which did not leave physical markers on the skeleton.

Bearing in mind the imponderable social and ideological codes which govern the relationships between the living and the dead, and the burial and evocation of the latter, much of the above discussion is, per force, speculative. There is enough tentative evidence at this stage, however, to show that social differentiation did exist in the late Mesolithic in most parts of Europe, and that this may have served as an additional variable in the determination of health and pathology among these populations.

Proposition 6: Neither skeletal morphology, nor genetic information derived from current European populations gives clear support to the colonisation model for the transition to farming.

We would argue that the skeletal morphology of Neolithic populations shows a continuation of trends that were already present in the Mesolithic. This is the case with changes in dental size and and in the reduction in stature, referred to above. The significance of this observation is that trends that are already visible in the Mesolithic cannot be used as evidence for the effects of the Neolithic "revolution".

As a regional sample, cranial materials from the Portuguese Neolithic site of Gruta do Zambujal, Melides, have been compared to Mesolithic samples from the Muge midden sites of Moita do Sebestiao and Cabeco da Arruda. No effective discrimination of the temporal samples was possible, and clustering techniques produced no site-centered groupings (Meiklejohn and Schentag, nd). Parallel findings have been made for the south of France by Duday (1977), though sample size is a problem in the latter case.

Congruent with such cases as the above, it can be demonstrated that there is change in form within the European Mesolithic. Regional populations of the Mesolithic are not uniform in appearance, the differences becoming more marked in late Mesolithic times. Changes between the Mesolithic and Neolithic populations have to be assessed against such a heterogenous background.

In summary, changes noted between Mesolithic and Neolithic cannot be used to support a model of population replacement if it has either been shown that such change has already been under way in the Mesolithic, or if the type of change is equivalent to differences already present between Mesolithic populations. The demonstration of population change at the Mesolithic-Neolithic boundary requires evidence for a modification in the trajectory of change, or evidence for disjunction in the overall pattern of morphology between replaced and replacing groups. A case for such morphological disjunction between Mesolithic and Neolithic populations has been made by Vencl (1986) for central, and by Menk and Nemeskeri (1989) for south-east Europe. However, there is an alternative explanation, with such change being seen as the result of a dietary shift to a softer diet, higher in fish and/or carbohydrate resources, rather than as a result of population replacement (y'Edynak, 1978, 1989; Zvelebil and Zvelebil, 1988).

Genetic patterning of extant populations has been used by Ammerman and Cavalli-Sforza (1984) to support a colonisation model for the diffusion of farming in Europe, referred to as demic diffusion. Aspects of the model have been both supported and criticised by archaeologists, physical anthropologists and geneticists alike. We find it instructive that there is a tendency for workers to criticise aspects of the study relevant to their own field of expertise while accepting those aspects in fields other than their own. Thus the archaeological aspects of the volume have received most support from physical anthropologists and geneticists but not from archaeologists and vice versa. We find the following points of particular interest with regard to the genetic evidence:

1) The diffusion model does not derive directly from the raw blood group and protein polymorphism data but from the Principal Components derived from the data. As Morton (1985) points out, both the mathematical basis of the components and their interpretation is heuristic. Thus though support for the diffusion model can be found in the Principal Component analysis, the hypothesis does not necessarily derive from the analysis.

2) Both Mourant (1985) and Livingstone (1985) point out that a key assumption of the model, including interpretation of the Principal Components, is that Rh negative (homozygotes for the cde allele complex) Basques represent a pre-Neolithic survival, later replaced by Rh positive populations from the east. However, the origin of this hypothesis is not based on evidence of the antiquity of the Basques, but on their Rh frequencies, thus creating a central circularity within the argument. Livingstone further points out that high cde frequencies are also seen in other alpine groups, suggestive of a positive selection for this allele complex under this environmental circumstance. Furthermore, Bantu expansion in sub-Saharan Africa, a much better documented and recent event, is not mirrored by the genetic variation data.

3) A final point that needs raising is that the Principal Components analysis suggests that the initial agricultural colonisation has been the most important population movement to affect Europe, since the First Component, constituting 27% of the total variation, is linked to this event. The implication is that events such as the Bronze and Iron Age expansions, the Roman hegemony and the Migration period are all of less significance in the formation of the current European genetic makeup. This may be so, but it leaves room for thought.

Given the above, we believe that the human biological evidence for demic diffusion as the explanation for agricultural origins is, at best, inconclusive (see also Sokal *et al*, 1990).

The case for the colonisation of Europe by Near Eastern farmers rests to a large extent on the identification of a dietary shift from foraging to farming with the arrival of the putative new population. Osteological data presented above, such as the rise in caries in the Mediterranean late Mesolithic, shows that dietary change is already occurring in some pre-Neolithic contexts. This reflects a cultural shift in subsistence pattern, detected archaeologically, towards greater use of vegetal foods (and also aquatic resources) in the European late Mesolithic.

Finally, with the possible exception of parts of central and south-eastern Europe, the evidence for continuity in material culture and settlement pattern argues against the applicability of the colonisation model beyond the European continental core (Zvelebil, 1986b; Zvelebil and Zvelebil, 1988).

Discussion

The Indigenous and Colonist Neolithisation in Europe

In our view, the biological data outlined above add to other evidence to show that:

1) The assumptions made by the colonisation or demic diffusion model may not be fully warranted, and may not necessarily follow from the evidence presented for them.

2) The model does not account for the transition to farming in areas of Europe other than possibly the eastern Mediterranean, southern Italy

and the Danube/Rhine valley system. Outside of this continental core, it would appear likely that farming was adopted by resident Mesolithic populations (Dennell, 1983; Barker, 1985; Zvelebil, 1986b; Price, 1987; Zvelebil and Zvelebil, 1988).The biological evidence is consonant with the notion of a slow and gradual transition to farming in most areas of Europe, a process which was not completed until the late Neolithic: a suggestion made by one of us earlier on the strength of cultural evidence alone (Zvelebil, 1986a,b).

Such conclusions shift our attention to the condition of the indigenous communities of the Mesolithic. Here we have to modify our first proposition, concerning the good health status of Mesolithic populations. Clearly, a dual pattern of pathology is emerging: a "west Mediterranean" zone, marked by the high incidence of caries and the relative absence of other insults, and the other, "north-east European" one, marked by the occurence of indicators of nutrient deficiency and by the absence of caries (Fig. 1).

Late Mesolithic: The West Mediterranean Pattern

The west Mediterranean pattern, marked by caries, is coeval with an increased intake of starchy carbohydrate foods, resurrecting the question, posed 15 years ago by Clarke (1976) of plant husbandry in the Mesolithic of this area (this pattern may appear early in the Mesolithic as seen from the evidence of sites such as Uzzo (Borgognini-Tarli *et al*, 1989; Constantini, 1989) and El Cingle Vermell (Turbon, 1986, 1989). Yet, comparative data from known farming contexts do not replicate this west Mediterranean pattern. The Levantine situation shows a lesser incidence of caries in the Natufian and the Neolithic, but a greater incidence of other stress-related insults, such as cribra orbitalia, porotic hyperostosis, enamel hypoplasia and osteoporosis (Angel, 1984; Smith *et al*, 1984, 1989), showing that, on the whole, the Levantine populations were less healthy than the west Mediterranean ones. Native North American cultivators suffered from high levels of both caries and other indicators of nutritional stress (Cassidy, 1984a,b; Cook, 1984; Goodman *et al*, 1984a; Larsen, 1984; Perzigian *et al*, 1984; Rose *et al*, 1984) with the distinction that, unlike the west Mediterranean pattern, caries also occured among the children. This suggests that soft carbohydrate diet was fed to juveniles in North America, a pattern that seems to be absent in Europe (Meiklejohn *et al*, 1988) (at the sites of Moita do Sebastiao and Cabeco da Arruda there are no cases where caries can be demonstrated prior to the eruption of the third molars at ca 18 years of age). This may reflect the use of pottery among the North American late Woodland and Mississippian groups for preparing soft meals (Buikstra *et al*, 1986), an option that was not available to European late Mesolithic groups until the 6th millennium bc. In summary, the west Mediterranean pattern would suggest to us early stages in the intensive use of plant foods, as yet not attended by markers of stress and reduced health which are common among many prehistoric agriculturists (e.g. Cohen and Armelagos, 1984b).

Late Mesolithic: The North and East European Pattern

If the pathology pattern observed at Iron Gates and in the south Baltic is representative of other hunter-fisher groups in northern and eastern Europe, then these late Mesolithic communities were worse off in terms of health than were their west Mediteraneran counterparts. This apparent paradox may pose a challenge for those who view forager groups dependant on aquatic resources as "affluent"(i.e. Wanatabe, 1968; Suttles, 1968; Akazawa, 1986; Hayden *et al*, 1985). However, there may be good reasons for the higher incidence of stress among the hunter-fisher groups in northern and eastern Europe. Even though, at sites such as Vlasac, there is evidence for the use of plant foods (Printz, 1987), absence of caries, and a high incidence of calculus and enamel hypoplasia suggest a protein dominated diet and recurring nutrient stress. Lack of a regular carbohydrate diet may have been a contributory factor in the failure to prevent periodic shortages arising from seasonal fluctuation in resources in the European continental interior.

But there is an alternative, and in our view more convincing explanation: that increased residential permanence and dependance on a fish diet resulted in parasitic infestations and infections which predjudiced the health of these hunter-fisher communities and resulted in the observed range of pathological conditions. Evidence from a number of wide-ranging contexts outside Europe supports this hypothesis: absence of caries, high incidence of rickets, osteomalacia, and porotic hyperostosis marked the pattern of pathology seen, for instance among the fish-eating community at Ra's al-Hamra along the east Arabian coast (Macchiarelli, 1989). Kent (1986) notes high incidence of porotic hyperostosis among the Eskimo and among the prehistoric marine-dependant Californian Indians. Increased sedentism and dependance on aquatic resources is a common denominator for all these groups.

How much stress is represented by these conditions? How far can stress be seen as an incentive for the adoption of farming? There are widely divergent views on this matter. In reference to the Levant, Smith *et al* (1984) commented:

> "The initial steps in plant and animal domestication were not associated with

environmental stress or deteriorating health status".

A similar position has been adopted by a number of other researchers (i.e. Roosevelt, 1984; Buikstra *et al*, 1986), while other workers maintain that population pressure and resource stress were the principal causes underlying the agricultural transition (i.e. Cohen, 1977; Binford, 1983; Redding, 1988).

The Demography of the Mesolithic-Neolithic Transition

Demographic parameters should play a major role in this discussion, since survival and reproduction are ultimate measures of fitness and a condition for social, as well as biological, reproduction. Bocquet-Appel and Masset (1982) and Johansson and Horowitz (1986) have outlined problems inherent in the palaeodemographic approach. Such problems (representativeness of the sample, assumptions of closure and constant growth rate, etc.: see Grauer, this volume) have led, for example, to three different estimates of life expectancy at Vlasac (Nemeskeri, 1978; y'Edynak, 1989). Given this situation, it follows that any palaeodemographic assessment has to be treated with caution.

However, at a very general level, there seems to have been a slow increase in life expectancy throughout the Palaeolithic to reach the value of 30-34 years in the late Upper Palaeolithic and Mesolithic of Europe and the Near East (Constandse-Westermann *et al*, 1984; Angel, 1984). In the Neolithic, life expectancy values remained stable, or have declined slightly in Europe and the Near East (Nemeskeri, 1978; Angel, 1984), while in North America life expectancy of agricultural populations is often reduced dramatically compared to hunter-gatherers and the transitional hunting-cultivating communities (Johansson and Horowitz, 1986; Buikstra *et al*, 1986; Eisenberg, this volume). At the same time, reproductive rates of prehistoric farmers remained generally unaffected and there is evidence of population increase. This points to the observation made by Rindos (1985): increased fitness is not always the same as adaptiveness. With the transition to farming, increased fitness, in terms of reproduction, has been attained at the expense of adaptation: that is, of equilibrium between the population and the environment, resulting in decline in life expectancy, decrease in stature and increase in stress during the Neolithic. It is only in extreme cases, when stress becomes acute and threatens the survival of a community that fitness will decline with maladaptation, as was the case with the Norse settlement in Greenland, or with the Averbuch Mississippian community in North America (see Eisenberg, this volume).

Against this background, it seems that levels of stress in the Mesolithic did not reach the threshold which would threaten the survival of the communities in question or cause a decline in life expectancy. Even at Vlasac, Nemeskeri and Lengyel (1978) accord the population "a rather good health status", while its uncorrected life expectancy is estimated at 36 years (Nemeskeri [1978]: other estimates range from 26 to 44 years, see also y'Edynak [1989]). Little change in these parameters can be seen as a result of the transition to the Neolithic: on the present evidence, a slightly higher evidence for pathology in the Mesolithic (see above) is compensated for by the slightly lower life expectancy estimates in the Neolithic. This would support the notion of a gradual transition to farming in Europe.

While these considerations support the view that levels of pathology and stress in the Mesolithic were not detrimental to the survival of a community, they do not exclude the possibility that the levels of stress, such as they were, served to encourage the adoption of farming. In seeking to understand this process, we again have to distinguish between the different patterns of the transition in the south-west and in the northern and eastern parts of Europe.

The Agricultural Transition in the West Mediterranean

In the west Mediterranean, the first elements of agro-pastoral "Neolithic" farming were adopted relatively early, at about 6000 bc, but the adoption was selective and staggered over two millennia, in that different elements of the system: caprines, bovines, cereals, pulses, ceramics and farming technology were adopted separately and gradually across the region (Guilaine, 1987; Vaquer, 1989; Binder, 1989; Lewthwaite, 1986; Zvelebil and Zvelebil, 1988). In other words, while the spread of some Neolithic traits was rapid throughout the Mediterranean (Lewthwaite, 1986), the substitution of the local by the Neolithic system was long. In our view, this pattern is consonant with the notion that husbandry practices were not really new: the adoption of neolithic cultigens and domesticates represented a continuation of practices that were already present, in the form of arboriculture of figs, olives and nut-bearing trees, and perhaps in the form of cultivation of small-seeded grasses and pulses. Given this situation, and bearing in mind the low levels of stress noted in the West Mediterranean, we do not believe that adoption of farming would confer an advantage on health grounds alone. While levels of stress detected in Mesolithic populations may have been a contributing factor in the adoption of farming, other more immediate causes must be sought within the socio- economic organisation at a regional level.

The Agricultural Transition in Northern and Eastern Europe

In the south Baltic and the Iron Gates, the situation was different. There, agro-pastoral farming was adopted relatively late by hunter-fisher groups, after it had been available in adjacent regions for about 1000 years (Zvelebil and Rowley-Conwy, 1984, 1986; Kozlowski and Kozlowski, 1986). During this "availability phase", exchange contacts developed between foragers and farmers across the agricultural frontier (Blankholm, 1987; Voytek and Tringham, 1989; Chapman, 1989). When the actual transition ocurred (ca. 3100 bc in Denmark, ca. 4500 bc in the Iron Gorges), it occured rapidly, within a period of perhaps 200 years. In other words, and in contrast to the west Mediterranean, the availability phase was long, but the substitution phase was short (Zvelebil and Rowley-Conwy, 1984, 1986). This was the case despite more evidence for stress present in these communnities.

Was this due to techological and social investment in the specialised, aquatically-oriented economy (Akazawa 1986)? Was this because exchange of products with the farmers, or perhaps, in the case of the Iron Gates, limited plant husbandry, may have gone some way to buffer these communities against nutritional stress? Y'Edynak (1989) notes some decline in enamel hypoplasia in the later levels at Vlasac, and links this with greater availability of plant food (see also Printz, 1987). Both of these factors may have played a part - but the delay in the adoption of farming, while stress was in evidence, does not support the view that residual stress such as was observed in eastern and northern Europe can provide the main reason for the agricultural transition.

Inevitably, such a conclusion is bound to shift attention to social factors. Social competition as the alternative explanation for the transition to farming has been advanced by Madsen (1982) and Jennbert (1984) for southern Scandinavia, by Bender (1985) for Brittany and by Lewthwaite (1986) for the west Mediterranean.

In this paper, we have argued for increase in social differentiation, sedentism and population density in the late Mesolithic in Europe, in particular among the hunter-fisher communities of northern and eastern Europe. Social competition, in whatever form, may have been given an added impetus by a long tradition of exchange in exotic goods across the agricultural frontier, the evidence for which can be seen in ceramics and stone artefact imports among the hunter-fisher communities of Scandinavia and the Iron Gates region. It should, finally, be noted that long distance exchange in exotic goods can also be demonstrated amoung the Mesolithic communities themselves (Constandse-Westermann and Newell, 1988).

Conclusion

In summary, our survey of pathology in European osteological samples related to the transition to farming revealed unexpected variability in their health status, related to different patterns in their diet and lifestyle. On the whole, we feel that the levels of stress observed were not sufficient to regard this stress as a direct cause of the agricultural transition. Immediate causes of the transition must be contingent on regional socio-economic and ecological conditions. In the west Mediterranean, the adoption of Neolithic cultigens and domesticates can be seen as a development of patterns of husbandry already present in the Mesolithic (ie. Clarke 1976). In northern and eastern parts of Europe, social competition, fostered by the proximity of farming communities, may have provided an immediate cause for the agricultural transition.

Social and ecological explanations for the transition to farming have been often treated as mutually incompatible, but this does not need to be the case. Both social and ecological conditions may have acted in combination to bring about the adoption of farming; since in both cases the range of solutions to problems caused by these conditions were located within the hunter-gatherer societies. Future research should examine social models for the transition to farming within this framework.

The biocultural study of human skeletal remains, as argued for here, is still in its early stages for European Stone Age populations. The inferences from the body of data currently available are compromised by small sample sizes, lack of representativeness, and a limited range of analysis. Rarely do the cultural and pathological/demographical analyses of the skeletal populations coincide. Thus for instance, the number of burial contexts analysed for pathology in the west Mediterranean is but a small sample of those in existence (Turbon, 1989:fig.1; Garralda, 1989). Or, the analysis of mortuary variation at Vlasac, and the relationship between social status and pathology at Vlasac would be of great importance for social models of the agricultural transition referred to above. Moreover, Vlasac is not the only skeletal population in the Iron Gates, although it is the only one comprehensively analysed (Garashanin, 1978). Full analysis of the skeletal remains from Lepenski Vir, Cuina Turcului, and Padina would add crucial information to our understanding of the region. Similarly, the analysis of the mortuary variability at Oleneostrovski Mogilnik (O'Shea and Zvelebil, 1984), the largest Mesolithic site in Europe, has not been matched, to date, by palaeopathological studies, at least not in published form. The paucity of fully analysed Neolithic samples has already been mentioned. Finally, the comparability of different case studies is adversely affected by variation in approach, criteria used in identification, and an all too frequent lack of attention to contextual information, both cultural and environmental (the Mesolithic of western and central Europe is the only period to have had a comprehensive evaluation of the context of attributed skeletal remains [Newell *et al*, 1979]). We have chosen to ignore these problems in this study, in the hope that significant patterns may become apparent nevertheless. Due to these problems, however, our conclusions must of necessity be preliminary, and will probably change as new evidence comes to light.

We feel than many of the problems outlined above can be overcome within the integrated, contextual, biocultural approach. In this study, we have tried to combine the existing biological data arising from the study of human skeletal remains with the cultural evidence. Processes of recovery and extrapolation from modern case studies introduce bias and uncertainty in both sets of data (Constandse-Westermann *et al*, 1984; Meiklejohn *et al*, 1984; Johansson and Horowitz, 1986). The relationship between the two is often subtle and complicated. Yet the meaning of each is much enhanced by the reference to the other. If this paper has succeeded in demonstrating the potential and the value of the biocultural approach, then it will have served its purpose.

References

Akazawa, T. (1986). Hunter-gatherer adaptions and the transition to food production in Japan, In *Hunters in Transition*. ed. M. Zvelebil. pp. 151-166. Cambridge: Cambridge University Press.

Alexandersen, V. (1988). Description of the human dentitions from the Late Mesolithic grave-fields at Skateholm, Sweden. In *The Skateholm Project. I. Man and Environment*. ed. L. Larsson. pp. 106-163. Stockholm: Almqvist and Wiksell.

Alexandersen, V. (1988). The Late-Mesolithic dentition in southern Scandinavia. *Rivista di Antropologia*, 66 (Supplement), pp. 191-204.

Ammerman, A.J. and Cavalli-Sforza, L.L. (1984). *The Neolithic Transition and the Genetics of Populations in Europe*. Princeton: Princeton University Press.

Andersen, S.H. (1987). Tybrind Vig: A submerged Ertebølle settlement in Denmark. In *European Wetlands in Prehistory*. eds. J.M. Coles and A.J. Lawson. pp. 253-280. Oxford: Clarendon Press.

Anderson, R.M. and May, R.M. (1979). Population biology of infectious diseases. *Nature*, 280:361-367.

Angel, J.L. (1984). Health as a crucial factor in the changes from hunting to developed farming in the Eastern Mediterranean. In *Paleopathology at the Origins of Agriculture*. eds. M.N. Cohen and G.J. Armelagos. pp. 51-73. Orlando: Academic Press.

Barker, G. (1985). Prehistoric farming in Europe. Cambridge: Cambridge University Press.

Bender, B. (1978). Gatherer-hunter to farmer: a social perspective. *World Archaeol.* 10(2):203-222

Bender, B. (1985). Emergent tribal formations in the American Midcontinent. *Am. Antiq.* 50(1):52-62.

Binder, D. (1989). Aspects de la Néolithisation dans les Aires Padane, Provencale et Ligure. In *Neolithisations* eds. O. Aurenche and J. Cauvin. pp. 199-225. British Archaeological Reports International Series 516.

Binford, L.R. (1968). Post-Pleistocene adaptations. In *New Perspectives in Archaeology*. eds. S. Binford and L.R. Binford. pp. 313-341. Chicago: Aldine.

Binford L.R. (1983). *In Pursuit of the Past*. London: Thames and Hudson.

Blankholm, H.P. (1987). Late Mesolithic hunter-gatherers and the transition to farming in southern Scandinavia. In *Mesolithic Northwest Europe: Recent Trends*. eds. P. Rowley-Conwy, M. Zvelebil & H.P. Blankholm. pp. 155-162. Sheffield: Department of Archaeology and Prehistory.

Bocquet-Appel, J.P. and Masset, C. (1982). Farewell to palaeodemography. *J. Hum. Evol.* 11:321-333.

Bogitsh, B.J. and Cheng, T.C. (1990). *Human Parasitology*. Philadelphia: Saunders College Publishing.

Bolomey, A. (1973). An Outline of late Epipalaeolithic Economy at the "Iron Gates": The Evidence on Bones. *Dacia* 17:1-12.

Borgognini-Tarli, S.M. and Repetto, E. (1985). Diet, dental features and oral pathology in the Mesolithic samples from Uzzo and Molara caves (Sicily). In *Papers in Italian Archaeology IV (II). Prehistory*. eds. C. Malone and S. Stoddart. pp. 87-100. British Archaeological Reports International Series 244.

Borgognini-Tarli, S.M. Della Santina, D. Francalacci, P. & Repetto, E. (1989). Reconstruction of Mesolithic diet using dental microwear and trace element analysis. The case of Grotta dell'Uzzo (Sicily). In *People and Culture in Change*. ed. I. Hershkovitz. pp. 283-320. British Archaeological Reports International Series, 508.

Brinch Petersen, E. (1973). A survey of late Palaeolithic Denmark. In *The Mesolithic in Europe*. ed. S.K. Kozlowski. pp. 77-127. Warsaw: Warsaw University Press.

Brinch Petersen, E. (1981). *Blev de myrdet alle tre?* Copenhagen: Munksgaard. 31 pp.. [*Stenalder et jaegerfolk for 7000 år siden*, 3]

Buikstra, J.E. Konigsberg, L.W. and Bullington, J. (1986) Fertility and The Development of Agriculture in the Prehistoric Midwest. *Am. Antiq.* 51(3):528-546

Calcagno, J.M. and Gibson, K.R. (1988). Human dental reduction: natural selection or probable mutation effect. *Am. J. Phys. Anthropol.* 77(4):505-517.

Carcuimaru, M. (1978). L'Analyse pollinique des coprolithes de la station archaeologique de Vlasac. In *Vlasac: A Mesolithic Settlement in the Iron Gates*, vol. II. ed. M. Garashanin. pp. 31-34. Belgrade: Serbian Academy of Sciences and Arts.

Cassidy, C.M. (1984a). Nutrition and health in agriculturalists and hunter-gatherers. In *Nutritional Anthropology*. eds. N.W. Jerome, G.H. Pelto & R.F. Kandel. pp. 117-145. Pleasantville, New York: Redgrave Publishing Company.

Cassidy, C.M. (1984b). Skeletal evidence for prehistoric subsistence adaptation in the Central Ohio River Valley. In *Paleopathology at the Origins of Agriculture*. eds. M.N. Cohen and G.J. Armelagos. pp. 307-345. Orlando: Academic Press.

Chagnon, N.A. (1972). Tribal social organisation and genetic microdifferentiation. In *The Structure of Human Populations*. eds. G.A. Harrison and A.J.

Boyce. pp. 252-282. Oxford: Oxford University Press.

Champion, T., Gamble, C., Shennan, S. & Whittle, A. (1984). *Prehistoric Europe*. Orlando: Academic Press.

Chapman, J. (1989). Demographic Trends in Neothermal South-East Europe. In *The Mesolithic in Europe*. ed. C. Bonsall. pp. 500-515. Edinburgh: John Donald.

Childe, G.V. (1928). *The Most Ancient East: the Oriental prelude to European prehistory*. London: Kegan Paul, Trench, Trubner & Co. Ltd.

Childe, G.V. (1957). *The Dawn of European Civilisation*. Bungay, Suffolk: The Chauser Press.

Clark, J.G.D. (1952). *Prehistoric Europe: The Economic Basis*. London: Methuen.

Clark, J.G.D. (1976). *The Earlier Stone Age Settlement of Scandinavia*. Cambridge: University Printing House.

Clark, G.A. and Neeley. M. (1987). Social Differentiation in European Mesolithic Burial Data. In *Mesolithic Northwest Europe, Recent Trends*. eds. P. Rowley-Conwy, M. Zvelebil and H.P. Blankholm. pp.121-130. Sheffield: The Department of Archaeology and Prehistory, University of Sheffield.

Clarke, D.L. (1976). Mesolithic Europe: The economic basis. In *Problems in Economic and Social Archaeology*. eds. I. Sieveking, I.J. Longworth and K.E. Wilson. pps. 449-481. London: Duckworth.

Clason, A.T. (1980). Padina and Starcevo:Game, Fish and Cattle. *Palaeohistoria* XXII. pp. 142-174.

Cohen, M.N. (1977). *The Food Crisis in Prehistory: Overpopulation and the Origins of Agriculture*. New Haven: Yale University Press.

Cohen, M.N. (1989). *Health and the Rise of Civilization*. New Haven: Yale University Press.

Cohen, M.N. and Armelagos, G.J. (1984a). Paleopathology at the origins of agriculture: editors summation. In *Palaeopathology at the Origins of Agriculture*. eds. M.N. Cohen and G.J. Armelagos. pp. 585-601. Orlando: Academic Press.

Cohen, M.N. and Armelagos, G.J. eds., (1984b). *Paleopathology at the Origins of Agriculture*. Orlando: Academic Press.

Constandse-Westermann, T.S. and Newell, R.R. in collaboration with Meiklejohn, C. (1984). Human biological background of population dynamics in the Western European Mesolithic. *Proceedings of the Koninklijke Nederlands Akademie van Wetenschappen*, Series B, *87*(2):139-223.

Constandse-Westermann, T.S. and Newell, R.R. (1984). Mesolithic trauma: demographical and chronological trends in Western Europe. In *Proceedings of the 4th European Meeting of the Palaeopathology Association, Middleburg/Antwerpen 1982*. eds. G.T. Haneveld and W.R.K. Perizonius. pp. 70-76.

Constandse-Westermann, T.S. and Newell, R.R. (1988). Patterns of extraterritorial ornaments dispersion: an approach to the measurement of Mesolithic exogamy. *Rivista di Antropologia*, 66(Supplement):75-126.

Constandse-Westermann, T.S. and Newell, R.R. (1989). Limb Lateralization and Social Stratification in Western Mesolithic Societies. In *People and Culture in Change*. ed. I. Herkovitz. pp.405-434. British Archaeological Reports International Series 508(ii).

Constantini, L. (1989). Plant exploitation at Grotta dell'Uzzo, Sicily. In *Foraging and Farming*. eds. D.R. Harris and G.C. Hillman. pp. 197-206. London: Unwyn Hyman.

Cook, D.C. (1984). Subsistence and health in the Lower Illinois Valley: osteological evidence. In *Paleopathology at the Origins of Agriculture*. eds. M.N. Cohen and G.J. Armelagos. pp. 235-269. Orlando: Academic Press

Courtin, J. (1974). *Le Néolithique de la Provence*. Paris

Dennell R. (1983). *European Economic Prehistory*. Orlando: Academic Press.

Duday, H. (1977). Difficulties d'une approche anthropologique de la Neolithisation. In *Approche Ecologique de l'Homme Fossile*. H. Laville and J. Renault-Miskovsky. pp. 367-370. Paris: Universite Marie et Pierre Curie.

Frayer, D.W. (1989). Les Vivants et Les Morts. Les Sepultures Du Paleolithique Superieur En Italie et Leur Interpretation. In *People and Culture in Change*. ed. I. Hershkovitz. pp. 435-458. British Archaeological Reports, International Series 508(ii).

Garashanin, M. (1978). *Vlasac: A Mesolithic Settlement in the Iron Gates*. Beograd: Serbian Academy of Sciences and Arts.

Garralda, M.D. (1989). Les Populations Post-Paleolithiques D'Espagne: Morphologie, Culture et Ecologie. In *People and Culture in Change*. ed. I. Hershkovitz. pp. 505-516. British Archaeological Research International Series 508(ii).

Goodman, A.H., Lallo, J., Armelagos, G.J. & Rose, J.C. (1984a). Health changes at Dickson Mounds (A.D. 950-1300). In *Paleopathology at the Origins of Agriculture*. eds. M.N. Cohen and G.J. Armelagos. pp. 271-305. Orlando: Academic Press.

Goodman, A.H., Martin, D.L., Armelagos, G.J. & Clark, G. (1984b). Indications of stress from bone and teeth. In *Palaeopathology at the Origins of Agriculture*. eds. M.N. Cohen. and G.J. Armelagos. pp. 13-49. Orlando: Academic Press.

Guillaine, J. (1987). ed. *Premieres Communautes Paysannes en Mediterranee Occidental e*. C.N.R.S. Paris,

Hansen, J.M. and Renfrew, J.M. (1978). Palaeolithic - Neolithic seed remains at Franchthi cave. *Nature*. 271:349 - 52.

Hayden, B., Eldridge, M., Eldridge, A. and Cannon, A. (1985). Complex hunter-gatherers in interior British Columbia. In *Prehistoric Hunter-Gatherers. The Emergence of Cultural Complexity*. eds. T.D. Price and J.A. Brown. pp. 181-197. Orlando: Academic Press.

Hengen, O.P. (1971). Cribra orbitalia: pathogensis and probable etiology. *Homo* 22(1):57-76.

Hillson, S. (1986). *Teeth*. Cambridge: Cambridge University Press.

Jacobsen, T.W. (1976). 17000 Years of Greek Prehistory. *Sci. Am.*. 234:76-87.

Jennbert, K. (1984). *Den produtkiva gåvan. Tradition och innovation i Sydskandinavien for omkring 5300 år sedan*. Lund: University of Lund [Acta Archaeologica Ludensia. Series in 4°,16].

Johansson, S.R. and Horowitz, S. (1986). Estimating mortality in skeletal populations: influence of the growth rate on the interpretation of levels and trends during the transition to agriculture. *Am. J. Phys. Anthropol.* 71:233-250.

Kent, S. (1986). The influence of sedetism and aggregation on porotic hyperostosis and anaemia: a case study. *Man* 21:605-36.

Kent, S. (1989). *Farmers as Hunters*. Cambridge: Cambridge University Press.

Kozlowski, J.K. and Kozlowski, S.K. (1986). Foragers of Central Europe and their acculturation. In *Hunters in Transition*. ed. M. Zvelebil. pp. 95-108. Cambridge: Cambridge University Press.

Lallo, J., Armelagos, G.J. & Mensforth, R.P. (1977). The role of diet disease and physiology in the origin of porotic hyperostosis. *Hum. Biol.* 40:471-483.

Larsen, C.S. (1984). Health and disease in prehistoric Georgia: the transition to agriculture. In *Paleopathology at the Origins of Agriculture*. eds. M.N. Cohen and G.J. Armelagos. pp.367-392. Orlando: Academic Press.

Lazenby, R.A. and McCormack, P. (1985). Salmon and malnutrition on the Northwest Coast. *Curr. Anthropol.* 26(3):379-384.

Lewthwaite, J. (1986). The transition to food production: a Mediterranean perspective. In *Hunters in Transition: Mesolithic societies of temperate Eurasia and their transition to farming*. ed. M. Zvelebil. pp. 53-66. Cambridge: Cambridge University Press.

Livingstone, F. (1985). Review of Ammerman and Cavalli-Sforza, The Neolithic Transition and the Genetics of Populations in Europe. *Hum. Biol.* 57(3):481-483.

Loze, I. (1979). *Pozdnii Neolit I Rannaya Bronza Lubanskoii Nizmennosti*. Zinatne, Riga.

Lubbock, J. (1865). *Pre-historic Times, as Illustrated by Ancient Remains, and the Manners and Customs of Modern Savages*. London: Williams and Norgate.

Lubell, D. and Jackes, M.K. (1988). Portuguese Mesolithic-Neolithic subsistence and settlement. *Rivista di Antropologia*. 66 (supplement):231-248.

Lubell, D., Jackes, M.K. and Meiklejohn, C. (1989). Archaeology and human biology of the Mesolithic-Neolithic transition in southern Portugal: a preliminary report. In *The Mesolithic in Europe: Papers presented at the 3rd International Symposium, Edinburgh 1985*. ed. C. Bonsall. pp. 632-640. Edinburgh: John Donald.

Macchiarelli, R. (1989). Prehistoric "fish-eaters" along the Eastern Arabian coasts. *Amer. J. Phys. Anthropol.* 78:575-594.

Madsen, T. (1982). Settlement systems of early agricultural societies in East Jutland, Denmark: a regional study of change. *J. Anthropol. Archaeol.* 1(3):197-236.

Marquardt, W.C. and Demaree, R.S. Jr. (1985). *Parasitology*. New York: Macmillan.

May, R.M. and Anderson, R.M. (1979). Population biology of infectious disease. *Nature* 280:455-461.

Meiklejohn, C. (1974). Biological concomitants of a model of band society. In *International Conference on the Prehistory and Paleoecology of Western North American Arctic and Subarctic*. eds. S. Raymond and P. Schledermann. pp. 133-141. Calgary: University of Calgary.

Meiklejohn, C. (1977). Genetic differentiation and deme structure: considerations for an understanding of the Athapaskan/Algonkian continuum. In *Problems in the Prehistory of the North American Subarctic: the Athapaskan Question*. eds. J. Helmer, F.J. Kense & J. van Dyke. pp. 106-110. Calgary: University of Calgary.

Meiklejohn, C., Schentag, C.T. Venema, A. & Key, P. (1984). Socioeconomic change and patterns of pathology and variation in the Mesolithic and Neolithic of Western Europe:some suggestions. In *Palaeopathology at the Origins of Agriculture*. eds. M.N. Cohen and G.J. Armelagos. pp. 75-100. Orlando:Academic Press.

Meiklejohn, C. and Schentag, C.T. (1988). Dental size in two Portuguese Mesolithic series and their implications. *Am. J. Phys. Anthropol.* 75:249 [abstract].

Meiklejohn, C., Baldwin, J.H. & Schentag, C.T. (1988). Carries as a probable dietary marker in the Western European Mesolithic. In *Diet and Subsistence: Current Archaeological Perspectives. Proceedings of the 19th Chacmool Conference*. eds. B.V. Kennedy and G.M. LeMoine. pp. 273-279. Calgary: University of Calgary Press.

Meiklejohn, C. and Schentag, C.T. (nd). *The Human Cranial Material from the Neolithic Site of Gruta do Zambujal, Melides, Portugal: Preliminary Description and Comparison*. Unpublished manuscript.

Menk, R., and Nemeskeri, J. (1989). The Transition from Mesolithic to Early Neolithic in Southeastern and Eastern Europe. In *People and Culture in Change*. ed. I. Hershkovitz. pp. 531-540. British Archaeological Reports International Series 508(ii).

Mensforth, R.P., Lovejoy, C.O., Lallo, J.W. & Armelagos, G.J. (1978). The role of constitutional factors, diet and infectious disease in the etiology of porotic hyperostosis and periosteal reactions in prehistoric infants and children. *Med.Anthropol.* 2(1):1-59.

Milisauskas, S. (1978). *European Prehistory*. Orlando: Academic Press.

Morton, N. (1985). Review of Ammerman and Cavalli-Sforza, The Neolithic Transition and the Genetics of Populations in Europe. *Am. J. Hum. Gen.* 37(4):830-831.

Mourant, A.E. (1985). Review of Ammerman and Cavalli-Sforza, The Neolithic Transition and the Genetics of Populations in Europe. *Ann. Hum. Biol.* 12(6):572-573.

Nemeskeri, J. (1978). Demographic Structure of the Vlasac Epippaleolithic Population. In *Vlasac: A Mesolithic Settlement in the Iron Gates*. ed. M. Garashanin. pp. 97-134. Beograd: Serbian Academy of Sciences and Arts.

Nemeskeri, J. and Lengyel, I. (1978). The results of paleopathological examinations. In *Vlasac: A Mesolithic Settlement in the Iron Gates* Vol II. ed. M. Garashanin. pp. 231-260. Beograd: Serbian Academy of Sciences and Arts.

Newell, R.R. (1984). On the Mesolithic contribution to the social evolution of Western European society. In *European Social Evolution: Archaeological Perspectives*. ed. J. Bintliff. pp. 69-82. Bradford: Bradford University Press.

Newell, R.R. Constandse-Westermann, T.S. & Meiklejohn, C. (1979). The skeletal remains of Mesolithic Man in Western Europe: an evaluative catalogue. *J. Hum. Evol.* 8(1):1-228.

Noli, D. and Avery, G. (1988). Protein poisoning and coastal subsistence. *J. Archaeol. Sci.* 15:395-401.

O'Shea, J., and Zvelebil, M. (1984). Oleneostrovski Mogilnik: reconstructing the social and economic organisation of prehistoric foragers in Northern Russia. *J. Anthropol. Archaeol.* 3:1-40.

Odell, G.H. (1974). Review of Kozlowski, The Mesolithic in Europe. *Helinium*, 14:278-284.
Pap, I. (1989). People and Environment in the Neolithic Age in Hungary. In *People and Culture in Change*. ed. I. Hershkovitz. pp. 385-396. British Archaeological Reports International Series 508(ii).
Perzigian, A.J., Tench, P.A. & Braun, D.J. (1984). Prehistoric health in the Ohio River Valley. In *Paleopathology at the Origins of Agriculture*. eds. M.N. Cohen and G. J. Armelagos. pp. 347-366. Orlando: Academic Press.
Piggot, S. (1954). *The Neolithic Cultures of the British Isles*. Cambridge: Cambridge University Press.
Piggot, S. (1965). *Ancient Europe*. Edinburgh: Edinburgh University Press.
Price, T.D. (1985). Affluent foragers of mesolithic Southern Scandinavia. In *Prehistoric Hunter-Gatherers: The Emergence of Cultural Complexity*. eds. T.D. Price and J.A. Brown. pp.341-360. Orlando: Academic Press.
Price, T.D. (1987). The Mesolithic of Western Europe. *J.World Prehist.* 1:225-305.
Price, T.D. (1989). The reconstruction of mesolithic diets. In *The Mesolithic in Europe*. ed. C Bonsall. pp. 48-59. Edinburgh: John Donald.
Price, T.D., and Brown, J. (1985). *Prehistoric Hunter-Gatherers: the Emergence of Cultural Complexity*. Orlando: Academic Press.
Printz, B. (1987). *Mesolithic Adaptions on the Lower Danume: Vlasac and the Iron Gates Gorge*. British Archaeological Reports International Series 330.
Redding, R.W. (1988). A general explanation of subsistence change: from hunting and gathering to food production. *J. Anthropol. Archaeol.* 7(1):56-97.
Renfrew, C. (1987). *Archaeology and Language: The Puzzle of Indo-Europeans Origins*. London: Johnathan Cape.
Rindos, D. (1985). Darwinian selection, symbolic variation and the evolution of culture. *Curr. Anthropol.* 26(1):65-88.
Roosevelt, A.C. (1984). Population, health, and the evolution of subsistence: conclusions from the conference. In *Paleopathology at the Origins of Agriculture*. eds. M.N.Cohen and G.J. Armelagos. pp.559-583. Orlando: Academic Press.
Rose, J.C., Burnett, B.A., Nassany, M.S. & Blaeuer, M.W. (1984). Paleopathology and the origins of Maize agriculture in the Lower Mississippi Valley and Caddoan culture areas. In *Paleopathology at the Origins of Agriculture*. eds. M.N. Cohen and G.J. Armelagos. pp. 393-424. Orlando: Academic Press.
Rowley-Conwy, P. (1983). Sedentary hunters: the Ertebølle example. In *Hunter-Gatherer Economy in Prehistory*. ed. G.N. Bailey. pp. 111-126. Cambridge: Cambridge University Press.
Rowley-Conwy, P. (1984). The laziness of the short distance hunter: the origins of agriculture in western Denmark. *J. Anthropol. Archaeol.* 3:300-324.
Rowley-Conwy, P. (1985). The origins of agriculture in Denmark: a review of some theories. *J. Danish Archaeol.* 4:188-195.
Rowley-Conwy, P. and Zvelebil, M. (1989). Saving it for later: storage by prehistoric hunter-gatherers in Europe. In eds. P. Halstead and J. O'Shea. *Bad Year Economics*. pp.40-56 Cambridge: Cambridge University Press.
Rozoy, J.G. (1978). *Les Derniers Chasseurs*. Bulletin de la Societe Archaeologique Champenoise, 3 vols.
Smith, P. (1989). Paleonutrition and subsistence patterns in the Natufians. In *People and Culture in Change*. ed. I. Herskovitz. pp. 375-384 British Archaeological Reports International Series 508(ii).
Smith, P., Bar-Yosef, O. & Sillen, A. (1984). Archaeological and skeletal evidence for dietary change during the late Pleistocene/early Holocene in the Levant. In *Paleopathology at the Origins of Agriculture*. eds. M.N. Cohen and G.J. Armelagos. pp. 101-136. Orlando: Academic Press.
Sokal, R.R., Oden, N.L., Legendre, P., Fortin, M.-J., Kim, J., Thomson, B.A., Vaudor, A., Harding, R.M., & Barbujani, G. (1990). Genetics and language in European Populations. *Am. Nat.* 135(2):157-175.
Speth, J.D. (1989). Early hominid and scavenging: the role of meat as an energy source. *J. Hum. Evol.* 18:329-343.
Speth, J.D. (1990). Seasonality, resource stress and food sharing in so-called 'egalitarian' foraging societies. *J. Anthropol. Archaeol.* 9(2):148-188.
Suttles, W. (1968). Coping with abundance: subsistence on the Northwest Coast. In *Man the Hunter*. eds. R. B. Lee and I. Devore. pp. 56-68. New York: Aldine.
Tauber, H. (1981). 13C evidence for dietary habits of prehistoric man in Denmark. *Nature, 292*:332-333.
Tringham, R. (1973). *Hunters, Fishers and Farmers of Eastern Europe 6000-3000 BC*. London: Hutchinson University Libarary.
Turbon, D. (1986). Resti umani epipaleolitici di "El Cingle Vermell" (Barcelona, Espagna). *Archivio per l'Antropologia e la Etnologia*. 116:205-213.
Turbon, D. (1989). Epipaleolithic remains from the Northeast Iberian Peninsula. In *People and Culture in Change*. ed. I. Herskovitz. pp. 495-503 British Archaeological Reports, International Series 508.
Vaquer, J. (1989). Innovation et inertie dans le processus de neolithisation en Languedoc occidental. In *Neolithisations*. eds. O. Aurenche and J. Calvin. pp. 187-197. British Archaeological Reports International Series 516.
Vaquer, J., Geddes, D., Barbaza, M. & Erroux, J. (1986). Mesolithic plant exploitation at the Balma Abeurador (France). *Oxford J. Archaeol.* 5(1):1-18.
Vencl, S. (1986). The role of hunting-gathering populations in the transition to farming: a Central European perspective. In *Hunters in Transition*. ed. M. Zvelebil. pp. 43-52. Cambridge: Cambridge Univiversity Press.
Voytek, B.A. and Tringham, R. (1989). Rethinking the Mesolithic: the case of South-East Europe. In *The Mesolithic in Europe*. ed. C. Bonsall. pp. 492-499. Edinburgh: John Donald.
Wanatabe, K. (1968). Subsistence and ecology of northern food gatherers with special reference to the Ainu. In *Man the Hunter*. eds. R. Lee and I. Devore. pp. 69-77. New York: Aldine.
y`Edynak, G. (1978). Culture, diet and dental deduction in Mesolithic forager-fishers of Yugoslavia. *Curr. Anthropol.* 19(3):616-617.
y`Edynak, G. (1989). Yugoslav Mesolithic dental reduction. *Am. J. Phys. Anthropol.* 78:17-36.
Zvelebil, M. (1986a). *Hunters in Transition*. Cambridge: Cambridge University Press.
Zvelebil, M. (1986b). Mesolithic societies and the transition to farming: problems of time, scale and organisation. In *Hunters in Transition: Mesolithic societies of temperate Eurasia and their transition to farming*. ed. M. Zvelebil. pp. 167-188. Cambridge: Cambridge University Press.

Zvelebil, M. and Rowley-Conwy, P. (1984). Transition to farming in northern Europe: a hunter-gatherer perspective. *Norwegian Archaeol. Rev.* 17:104-28.
Zvelebil, M. and Rowley-Conwy, P. (1986). Foragers and farmers in Atlantic Europe. In *Hunters in Transition: Mesolithic Societies of Temperate Eurasia and Their Transition to Farming.* ed. M. Zvelebil. pp. 67-93. Cambridge: Cambridge University Press.
Zvelebil, M. and Zvelebil, K.V. (1988). Agricultural transition and Indo-European dispersals. *Antiquity.* 62:574-583.
Živanovic, S. (1975). A note on the anthropological characteristics of the Padina population. *Zeitschrift Morphol. Anthropol.* 66:161-175.

Acknowledgement

C. Meiklejohn would like to thank M. Zvelebil for the invitation to the TAG88 conference and the suggestion that this joint paper would be useful. Attendance at the Sheffield meting was made possible by a Research Furlough Grant from the University of Winnipeg. Aspects of the work reported were made possible by grants 410-87-1045 and 410-89-0376 to C. Meiklejohn by the Social Sciences and Humanities research Council of Canada.

TABLE 11.1

CARIES RATES IN THE EUROPEAN MESOLITHIC

Region/Site	Caries Number	Caries Rate [%]
Skateholm I, Sweden	5/848	0.59
Skateholm II, Sweden	0/491	0.0
Vedbaek-Bøgebakken, Denmark	0/357	0.0
Other non-cemetery materials	1/203	0.49
Total Scandinavia	**6/1899**	**0.315**
Téviec, France	8/352	2.27
Hoedic, France	7/133	5.26
Other non-cemetery materials	0/181	0.00
Total Mid-Latitude Europe	**15/666**	**2.25**
Moita do Sebastiao, Portugal	107/889	12.04
Cabeco da Arruda, Portugal	85/1176	7.23
Uzzo and Molara, Italy	28/223	12.56
El Cingle Vermell, Spain	8/9	88.88
Other non-cemetery materials	2/128	1.57
Total Mediterranean Europe	**230/2425**	**9.48**
TOTAL EUROPEAN MESOLITHIC	***251/4990***	***5.03***

Sources as in Meiklejohn et al. (1988) except:

Skateholm	Alexandersen, 1988
El Cingle Vermell	Turbon, 1986

These figures take precedence over those published in Meiklejohn, et al.(1988).

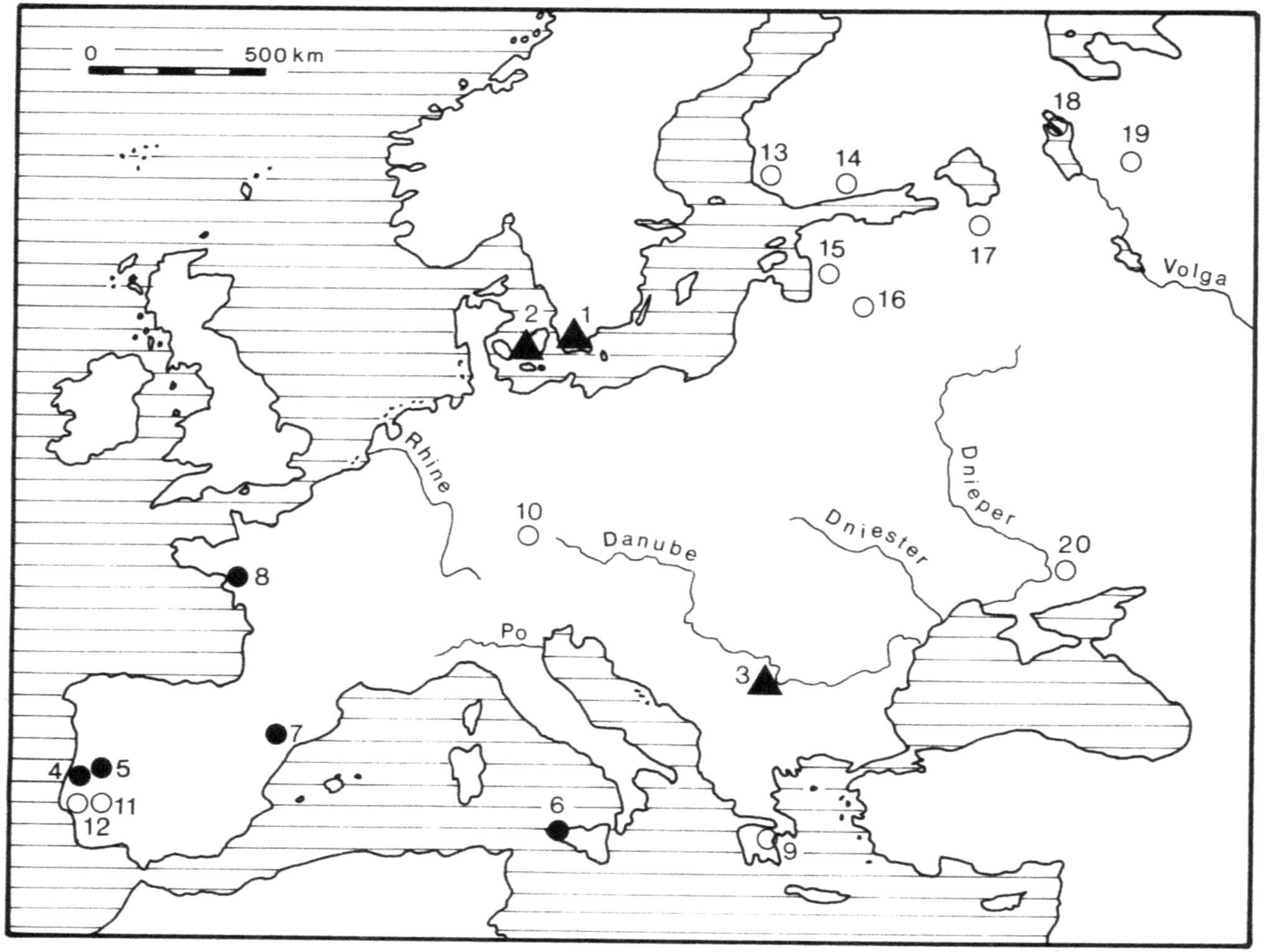

Sites characteristic of 'North-East European' pattern.

1. Skateholm, Sweden
2. Vedbaek, Denmark
3. Iron Gates Sites (Vlasac, Padina), Yugoslavia

Sites characteristic of the 'West Mediterranean' pattern

4. Moita do Sebestião, Portugal
5. Cabeço da Arruda, Portugal
6. Uzzo and Molara, Italy
7. El Cingle Vermell, Spain
8. Hoedic and Teviec, France

Other burial sties

9. Franchthi Cave, Greece
10. Ofnet, Germany
11. Cabeço do Pez, Portugal
12. Cabeço das Amoreiras, Portugal
13. Kolmhaara, Finland
14. Jonsas, Finland
15. Zveinieki, Latvia
16. Abora, Latvia
17. Sandermokha, Russia
18. Oleneostrovskii Mogilnik, Karelia
19. Popovo, Russia
20. Vasilevka 103, Ukraine

Figure 11.1 Cemeteries and Burials in Mesolithic Europe

www.ingramcontent.com/pod-product-compliance
Lightning Source LLC
LaVergne TN
LVHW071216210926
848482LV00015B/828
* 9 7 8 0 8 6 0 5 4 7 1 6 7 *